L. Gierl/A. D. Cliff/A.-J. Valleron/
P. Farrington/M. Bull (Eds.)

GEOMED '97

GEOMED '97

Proceedings of the International Workshop on Geomedical Systems

Rostock, Germany, September 1997

Edited by

Prof. Dr. Lothar Gierl, University of Rostock
Prof. Dr. Andrew D. Cliff, University of Cambridge
Prof. Dr. Alain-Jacques Valleron, Pierre-et-Marie-Curie University, Paris
Dr. Paddy Farrington, PHLS Communicable Disease Surveillance Centre, London
Dipl.-Math. Mathias Bull, University of Rostock

B.G.Teubner Stuttgart · Leipzig 1998

Professor Lothar Gierl was born in Munich, Germany, in 1942. He has studied communication engineering at the University of Technology of Munich. From 1978 to 1995, he was senior software engineer and lecturer at the computer centre of the University Hospital Großhadern, University of Munich. Since 1995, he is professor and director of the Institute for Medical Informatics and Biometry, University of Rostock.
Fields of Interest: medical information systems, expert systems, case-based reasoning.

Professor Andrew D. Cliff was born in Stamford, England, in 1943. He has held academic posts at the University of Bristol, England (1968 – 1971) and the University of Cambridge since 1972 where he is currently Professor of Theoretical Geography.
Fields of interest: spatial systems analysis, demography, geographical aspects of epidemiology.

Professor Alain-Jacques Valleron was born in Neuilly sur Seine, France, in 1943. He has graduated from Ecole Polytechnique, Paris and from the University of Paris. He is presently Professor of Public Health at the Université Pierre et Marie Curie and Director of a research Unit of INSERM (the French NTH) entitled "Epidemiologie et Sciences de l'Information".
Fields of interest: communicable diseases, iatrogenic risks, modelling in epidemiology.

Dr. Paddy Farrington was born in Paris, France, in 1954. He studied mathematics at the universities of Aberdeen, Oxford and Leeds. A latecomer to statistics, which he took up in 1986 after a few years indulging his other interests, he is currently deputy head of the Statistics Unit at the Public Health Laboratory Service in London, UK.
Fields of interest: statistical modelling, particularly generalised linear models and their extensions, and the applications of statistics to infectious diseases.

Dipl.-Math. Mathias Bull was born in Bad Doberan, Germany, in 1961. From 1981 to 1986, he has studied mathematics at the University of Greifswald.
Fields of interest: intelligent information systems.

Gedruckt auf chlorfrei gebleichtem Papier.

Die Deutsche Bibliothek – CIP-Einheitsaufnahme

GEOMED <1, 1997, Rostock>:
GEOMED '97 : proceedings of the International Workshop on Geomedical Systems, Rostock, Germany, September 1997 / ed. by Lothar Gierl ... – Stuttgart ; Leipzig : Teubner, 1998

ISBN 978-3-8154-2311-0 ISBN 978-3-322-95397-1 (eBook)
DOI 10.1007/978-3-322-95397-1

Foreword

A key event in the development of modern epidemiology was the discovery by the English physician, John Snow, that cholera is transmitted by contaminated water. During the cholera epidemic in London in 1854, Snow mapped the locations of cholera deaths, observed a cluster of victims in a particular neighbourhood and found that most of these cases had drunk water from a communal water pump. The handle of the pump was removed at Snow's insistence, and the epidemic ended within a few days.

Since these early days, the science of epidemiology has grown into a major discipline, with many successes to its credit. Many of the diseases which wreaked havoc in the last century have been brought under control, and in the case of smallpox, eliminated, through improvements in hygiene and the use of preventive and control measures such as mass vaccination.

Nevertheless, in recent years, new problems have emerged, and old diseases have re-emerged. Many foodborne and waterborne disease outbreaks go unrecognized or are detected too late for effective control measures to be implemented. New infections, such as HIV, present new threats. Antimicrobial drug resistance, particularly the increase in drug resistant TB, also poses new challenges.

The reasons for the emergence or re-emergence of these new infections are complex. Factors responsible include social changes such as mass population movements, rural-to-urban migration and accelerated urbanisation, population growth, rapid transport, new food technologies, and new life styles, as well as environmental changes which increase the risk of exposure to zoonotic or vector-borne infections, such as altered land use patterns and irrigation.

The multiplicity of factors involved suggests that to confront these problems effectively requires a broad, multidisciplinary approach. Policies to promote practical disease prevention measures must be based on up-to-date information on current health risks as well as on outbreaks and the spread of communicable diseases. In many cases, however, the information systems currently used to monitor human and animal infectious diseases domestically and internationally are inadequate.

Geomedical systems are information systems used to monitor health in geographically-distributed populations. They are used both to detect and to forecast health risks. In recent years, many national and international projects involving such systems have been established, often taking advantage of the wide availability of suitable software. When combined with formal methods of statistical analysis, these new informatics tools can fruitfully be applied to epidemiological problems, including the detection of outbreaks and the visualisation and investigation of disease patterns.

The figure below illustrates the links between geomedical systems, medicine, statistics, geography and computer science.

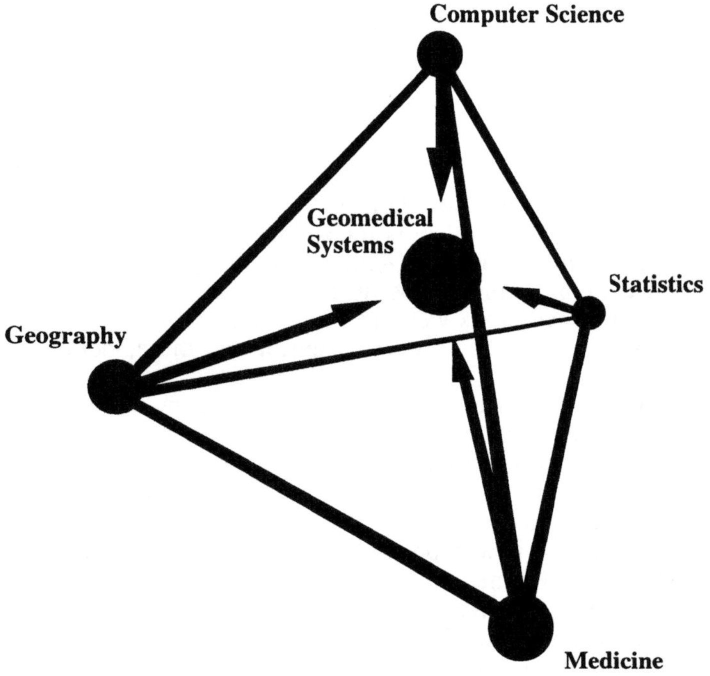

The primary aim of the GEOMED '97 workshop, which took place in Rostock in September 1997, was to provide a multidisciplinary focus on these issues, with the emphasis on geomedical systems, but also including contributions from the field of epidemiology, geography, and statistics.

The papers in the Proceedings of the GEOMED'97 Workshop present new directions in the field of geomedical systems and explore the interface with related topics such as epidemiology and statistical modelling. We thank all those who contributed to the success of the workshop.

The GEOMED'97 workshop took place from September 4 to September 6, 1997 in the *InterCityHotel* Rostock, whose hospitality we gratefully acknowledge. The Workshop was sponsored by Deutsche Forschungsgemeinschaft (German Science Foundation). The publication of the Proceedings is supported by the Medical Faculty of the University of Rostock, Merck KGaA, Darmstadt, and JANSSEN-CILAG GmbH, Neuss.

Programme Committee

H-J. Appelrath (Oldenburg, Germany)
A. D. Cliff (Cambridge, United Kingdom)
C. P. Farrington (London, United Kingdom)
S. M. Freire (Rio de Janeiro , Brazil)
L. Gierl (Rostock, Germany), chair
R.Thomas (Manchester, United Kingdom)
H. Uphoff (Marburg , Germany)
J.-A. Valleron (Paris, France)

Organisation Committee

A.Brauer, M. Bull (chair), G. Kundt,V. Müller,
R. Schmidt (Rostock, Germany)

Lothar Gierl, University of Rostock
Andrew David Cliff, University of Cambridge
Alain-Jacques Valleron, University of Paris
Paddy Farrington, PHLS London
Mathias Bull, University of Rostock

December 1997

Contents

Geographical Epidemiology

Statistical Methods

Systems

Geographical Epidemiology

Detecting Space–Time Patterns in Geocoded Disease Data. Cholera in London, 1854 Measles in the United States, 1962–95

A.D. Cliff [1], P. Haggett [2], & M.R. Smallman-Raynor [3]

[1] Department of Geography, University of Cambridge, Downing Place, Cambridge, CB2 3EN, England (e-mail: adc2@hermes.cam.ac.uk).

[2] Department of Geography, University of Bristol, Clifton, Bristol, BS8 1SS, England (e-mail: Karine.Taylor@bristol.ac.uk).

[3] Department of Geography, University of Nottingham, University Park, Nottingham, NG7 2RD (e-mail: raynor@geography.nottingham.ac.uk).

Abstract

This paper illustrates some of the methods used by geographers to specify appropriate models of the space-time patterns found in geocoded disease data. Such data are of two basic forms: point or area-based. Two data sets are analyzed. For point patterns, we use the classic data collected in 1854 by Dr. John Snow which give the geographical location of deaths from cholera in the Golden Square (Soho) district of London, England. Simple mapping methods based upon measures of geographical centrality (the spatial mean, median and mode) and Thiessen polygon techniques point to one particular water pump as the source of the cholera outbreak. The results of these methods are compared with those from nearest neighbour analysis as developed by Ripley (1976) and Diggle (1983). For area-based data, we use the monthly reported number of measles cases per million population in the United States between 1962 and 1995 at different spatial scales. The techniques of spatial autocorrelation analysis, disease centroids, and multidimensional scaling are used to unravel the distinctive geography of measles transmission into and within the United States that resulted from the systematic vaccination programmes articulated from 1967 by the Centers for Disease Control (CDC).

1 Introduction

The purpose of this paper is to illustrate some of the methods used by geographers to specify appropriate models of the space-time patterns found in geocoded disease data. Such data are of two basic forms: point or area-based. A spatial point pattern is a collection of points (ir)regularly located within a bounded region of space. The data set may consist of locations only, or it may be a *marked* point process, with data values (the marks) associated with each location. For area-based data, the values of variables are associated with a system of non-overlapping (ir)regular polygons that have been used to partition the region of space. The methods used to analyze area-based data can also generally be used to study a marked point process.

To give unity to the discussion, two data sets, each with its own intrinsic interest, are analyzed. For point patterns, we use the classic data collected in 1854 by Dr. John Snow which give the geographical location of deaths from cholera in the Golden Square (Soho) district of London, England. Simple mapping methods based upon measures of geographical centrality (the spatial mean, median and mode) and Thiessen polygon techniques point to one particular water pump as the source of the cholera outbreak. The results of these methods are compared with those from nearest neighbour analysis as developed by Ripley (1976) and Diggle (1983).

For area-based data, we use the monthly reported number of measles cases per million population in the United States between 1962 and 1995 at different spatial scales: the nation, Bureau of the Census regions, and the 48 states and Washington DC that comprise the conterminous US. The techniques of spatial autocorrelation analysis, disease centroids and multidimensional scaling are used to unravel the distinctive geography of measles transmission into and within the United States that resulted from the systematic vaccination programmes articulated from 1967 by the Centers for Disease Control (CDC). These programmes were designed to eliminate indigenous measles, a goal possibly achieved in 1996 (eliminating the disease is, of course, different to eradicating the virus). It is shown that inter-area measles transmission in epidemic episodes over the period of the campaigns was driven by two components of spread – first, from larger centres to smaller (hierarchical diffusion) and, second, between contiguous geographical units (contagious diffusion). The main population centres acted as disease reservoirs from which measles diffused by local spatial spread during epidemic episodes. Population effects gained at the expense of contagious diffusion as the campaign progressed, while the continued import of measles from international sources also assumed increasing significance.

2 Point Pattern Analysis: Cholera in London, 1854

2.1 The Data

In the nineteeth century, a series of six pandemics of Asiatic cholera swept the globe. The third of these great pandemics, lasting from 1840-55, played a classic role in the unravelling of the causes of cholera. Its impact upon Great Britain in 1849 provided John Snow with clues to the water-borne nature of the disease. It led in 1854 to one of medical history's most fascinating detective stories, a story which has been told and retold and which led to Snow's isolation of the mode of transmission of the disease.

The 1849 visitation of cholera to Great Britain resulted in over 52,000 deaths and affected nearly 500 settlements, and its course led Snow to comment that it was "since the latter part of 1848, when I first arrived at my present conclusions respecting the mode of communication of cholera" (Snow, 1854, p. 125). These ideas–that the disease is carried in water contaminated with faecal matter–were published in 1849 in a slim book, *On the Mode of the Communication of Cholera*. His work led on to a much enlarged second edition, published in 1854, which contained his celebrated map and account of the Broad Street pump affair.

2.2 The Affair of the Broad Street Pump

During a cholera epidemic which affected London in the late summer of 1854, Snow demonstrated that most of the cholera deaths in a restricted area of London near Golden Square, Soho, occurred among people who had drunk water from a single pump in Broad Street which had become contaminated by seepage from a leaking cesspool or drain. Of the conflagration, Snow himself wrote:

> The most terrible outbreak of cholera which ever occurred in this kingdom, is probably that which took place in Broad Street, Golden Square, and the adjoining streets, a few weeks ago. Within two hundred and fifty yards of the spot where Cambridge Street joins Broad Street, there were upwards of five hundred fatal attacks of cholera in ten days. The mortality in this limited area probably equals any that was ever caused in this country, even by the plague . . . (Snow, 1854, p. 38).

2.2.1 Spatial Mapping

As part of his investigation, Snow mapped the geographical distribution of cholera deaths in the vicinity of Golden Square. Snow's investigation can therefore be re-expressed as a geographical problem, that of deciding where the main centre of deaths was located. Snow's map is reproduced in Fig. 1A, and shows the locations of 571 deaths from cholera situated within a radius of about 250 metres of the Broad Street pump. The deaths are drawn as a point pattern with the underlying mesh of streets and dwellings omitted. Diagrams (B) and (C) are obtained directly from (A). In each, the numbers of deaths falling within a particular row or column of the lattice, as defined by the marginal ticks drawn round the perimeter of (A), are accumulated and plotted as histograms.

The centrality of a spatial distribution can be described by three measures based on the elementary statistical parameters of the mean, the median and the mode. The position of each parameter is marked on the two histograms, (B) and (C). The location of the geographical centre, as defined by a particular parameter, is fixed at the intersection of the positions of that parameter in the row and column histograms. These are shown by the symbols on map (A). The mean and the median are relatively stable measures and are located within a few yards of each other; the mode is much more heavily affected by the shape of the histograms. To emphasize this point, we have enlarged the central part of Snow's original map in (D). This shows that the locations of the spatial mean and median fall within 20 metres the offending pump. It is interesting to reflect that, had the measures of geographical centrality described here been available to Snow, they would have pointed unerringly to the probable source of the contaminated water.

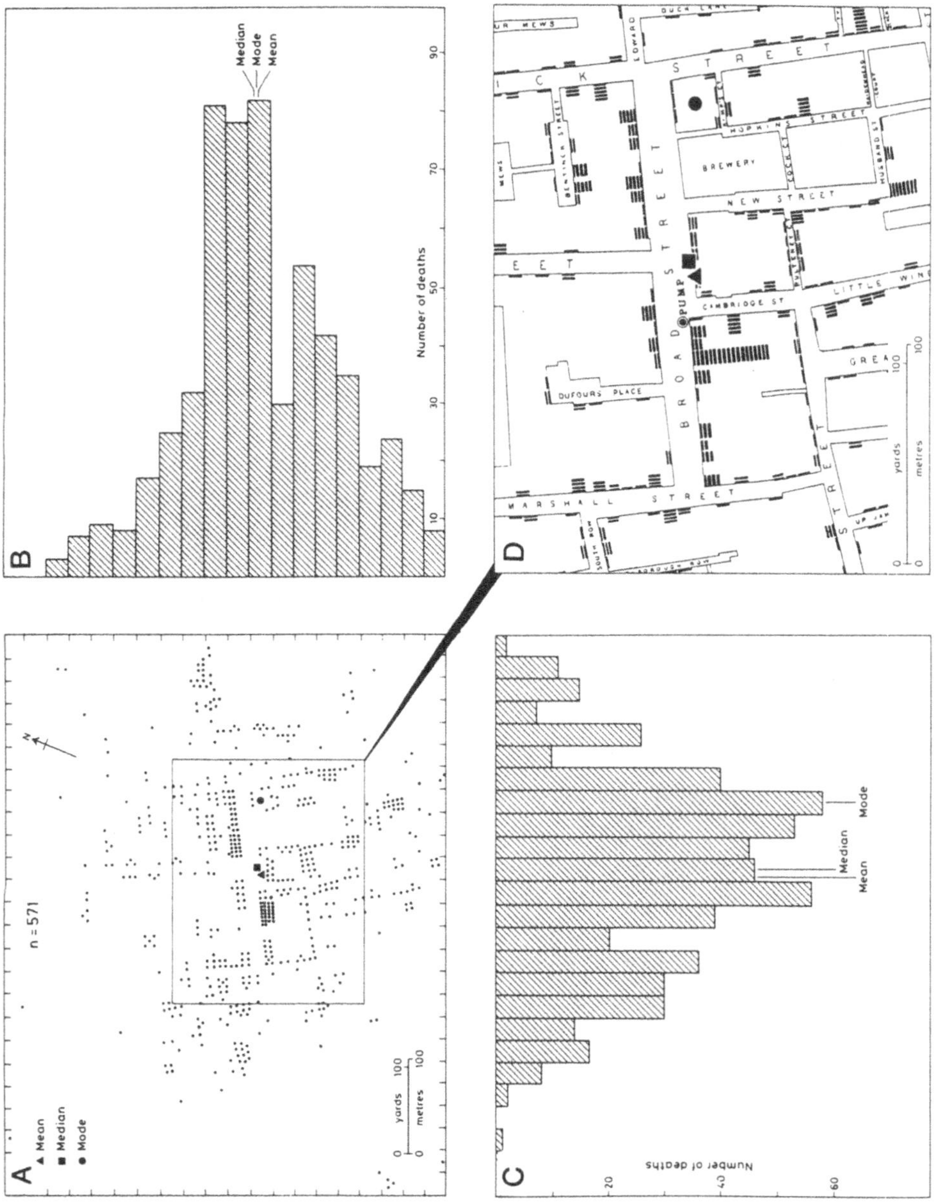

Fig. 1. Snow and cholera, London 1854. (A) Distribution of cholera deaths in Golden Square, London, as a point pattern. Histograms of north-south (B) and east-west (C) distributions of deaths are shown. (D) Central section of Snow's map of cholera deaths around the Broad Street pump. Source: Cliff and Haggett (1988, Fig. 1.15, p. 50).

2.2.2 Point Pattern Analysis

Nearest Neighbour Methods

The clustering of cholera deaths about the pump shown in Fig. 1A may be established more formally by nearest neighbour analysis. Following Diggle (1983, p. 4), we wish to compare the observed spatial point pattern with that which would occur under complete spatial randomness (CSR). We define CSR by the following criteria:

1. The intensity (number of points per unit area) of the point pattern does not vary over the region, A. Formally, the number of points in the planar region A with area $|A|$ follows a homogeneous Poisson distribution with mean $\lambda|A|$, where λ is the constant intensity.
2. There are no interactions among the points. Given n points with locations denoted by the vectors $\mathbf{x}$ in region A, the $\mathbf{x}$ are an independent random sample from the uniform distribution on A.

Define a nearest neighbour distance d_i as the distance from the i-th point to the nearest other point in A. The empirical distribution function, $\hat{G}$, say, of these point-to-point nearest neighbour distances can be used for comparison to a CSR process, where

$$\hat{G}_{(y)} = n^{-1} \sum_{d_i < y} 1, \tag{1}$$

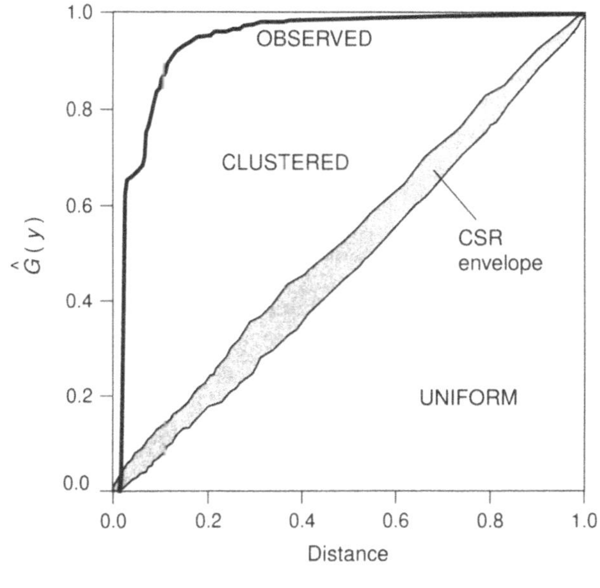

Fig. 2. Deaths from cholera, Soho, 1854. Empirical distribution function of point-to-point nearest neighbour distances with upper and lower simulation envelope from $s = 20$ realisations of a completely spatially random (CSR) process.

and n is the number of points in A. To test $\hat{G}$ for significance, taking into account edge effects for the region A, we can simulate the empirical distribution function of nearest neighbour distances for s realisations of a CSR process on A. The average of the s simulations provides a reference line, while the maximum and minimum provide a simulation envelope. If $\hat{G}$ calculated from the data falls outside this envelope, at either short or long nearest neighbour distances, there is evidence against CSR. Fig. 2 illustrates the results of this analysis, for $s = 20$, applied to Snow's data. For convenience, the cartesian coordinates of the deaths shown in Fig. 1A have been scaled onto a unit square. The plot shows strong evidence for clustering of the cholera deaths, with a massive excess of low-distance neighbours. Further analysis of the data using Ripley's (1976) *K-function* confirms this conclusion.

Thiessen Polygons

For many diseases, it is important to known the number of individuals who might potentially be exposed to a point source of infection – leukaemia clusters around nuclear power stations are an obvious example. The degree of accessibility of any geographical area or population to a point source may be resolved using the method of *Thiessen polygons.* Fig. 3A outlines the basis of the method using Snow's 1854 cholera data. This shows the distribution of water pumps in Soho in relation to the street pattern as recorded by Snow on map 1 of his monograph, *On the Mode of Communication of Cholera.* The location of the Broad Street pump is marked.

If we assume that people generally visited the geographically-nearest pump to obtain their drinking and cooking water (not unreasonable given the density of water) then, under Snow's hypothesis, cholera deaths should be greater within the catchment area of the Broad Street pump than in the catchment areas of the other pumps. These *spheres of influence* of the pumps may be delimited by Thiessen polygons.

The procedure for constructing the polygons is shown in diagram (B). Three steps are involved:

(a) Lines are drawn joining a given pump to each adjacent pump.

(b) Each of these inter-pump lines is bisected to give the midpoint of the line. The locations of these bisectors are shown by the ticks.

(c) From these midpoints, boundary lines are drawn at right angles to the original inter-pump lines to define a series of polygons.

The polygon around the Broad Street pump is shaded in diagram (B). Note that not all the midpoints on the inter-pump lines radiating from a particular pump are involved in this process. The basic property of any Thiessen polygon is that it contains all the area of any map which is geometrically nearer to its point centre (here a particular pump) than it is to any other point centre on the map. The complete set of polygons around all of the pumps given on Snow's map is shown in diagram (C). The number of deaths within each is also recorded using Snow's data as mapped in Fig. 1A. The excess within the catchment area of the Broad street pump (348 as compared to the next largest of 64) is apparent.

So far it has assumed that access to each pump was determined by 'crow-fly' distances. While the physical effort of carrying water meant that most people visited their nearest pump, cognisance should be taken of the complex street pattern in this area of London. Therefore in (D), the Thiessen polygons have been adjusted to take into account the patterns of access made possible by the street system shown in diagram (A). The modified polygons enclose areas which are closest to each pump in terms of street distance. The revised number of deaths within each has also been calculated. Again, the excess mortality in the catchment area of the Broad Street pump is evident. Simulation can be used to compare observed polygon counts with those arising from different spatial models.

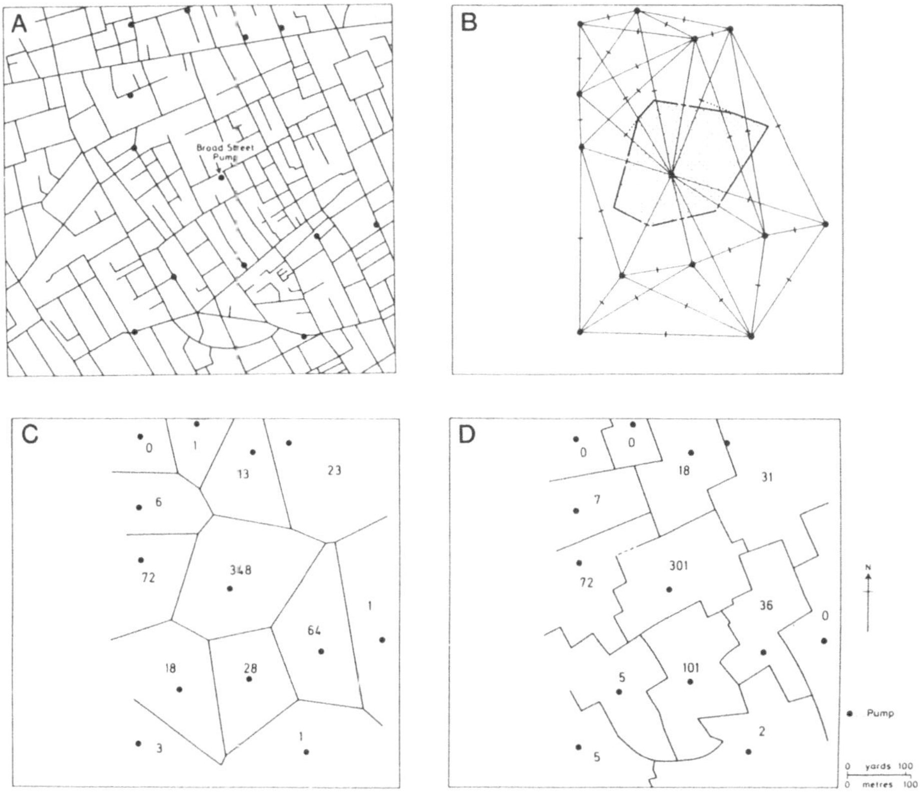

Fig. 3. Thiessen polygon analysis of Snow's cholera data. (A) Pump locations and street pattern, Soho 1854. (B) Polygon construction around each pump. (C) Thiessen polygons based upon crow-fly distances. (D) Thiessen polygons adjusted for Soho street pattern. Source: Cliff and Haggett (1988, Fig. 1.16, p. 54).

3 Area-based Data. I: Internal Spread of Measles in the United States, 1962–95

Most of the disease data collected by surveillance systems around the world relate to areas – world regions, countries, counties, cities, and so on – and it is to a consideration of appropriate geographical techniques for their analysis that we now turn. As noted in Section 1, our basic data set is the monthly time series of reported measles cases per million population, 1962-95, and associated vaccination data where available, for each of the 48 states and Washington DC that comprise the conterminous United States. From these data, regional and national time series may be constructed.

3.1 The Data

At the start of the twentieth century, thousands of deaths were caused by measles each year in the United States; even in the decade of the 1950s, an annual average of more than half a million measles cases and nearly 500 deaths was still being reported (Fig. 4). It was against this background that CDC evolved by 1966 a programme for the elimination of indigenous measles (Senser, Dull and Langmuir, 1967). The programme had four tactical elements: (a) routine immunization of infants at one year of age; (b) immunization at school entry of children not previously immunized; (c) surveillance; and (d) epidemic control. The immunization target aimed for was 90 percent of the childhood population.

Following the announcement of possible measles elimination, considerable effort was put into mass measles immunization programmes throughout the United States. Federal funds were appropriated and, over the next three years, an estimated 19.5 million doses of vaccine were administered (Cliff, Haggett and Smallman-Raynor, 1993, pp. 217-44). As Fig. 4 shows, this caused a significant reduction in the time series of reported cases, with the reported incidence plummeting in 1968 to around 5 percent of its 1962 value. Despite this dramatic success, it was only in 1996 that CDC argued that indigenous measles had been eliminated. Many reasons have been adduced for the length of the elimination campaign. These include the emergence in the US of new disease risks such as AIDS, Legionnaires' and Lyme which shifted the focus of attention, a general reduction in vaccination levels against measles as public awareness fell with the decline of the disease as a health problem, the inaccessibility of segments of the population to vaccination (certain religious groups, and ethnic minorities in inner cities, for example) and the continual import of measles into the US from overseas sources. The retreat of measles in the United States produced a distinctive geography which we illuminate in this and the next two sections.

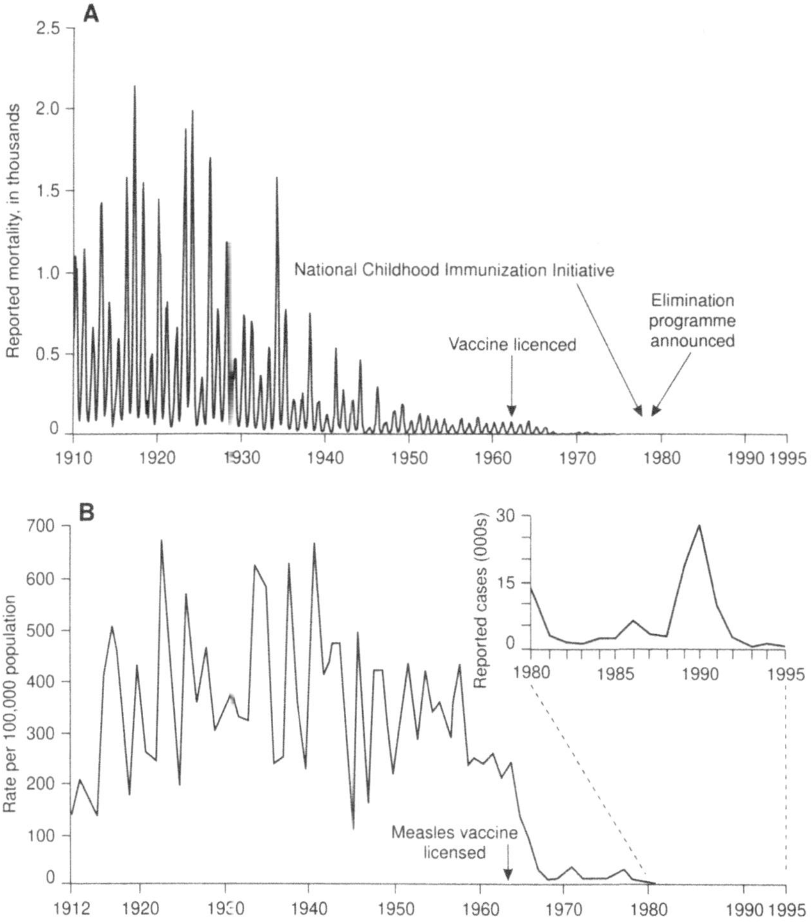

Fig. 4. United States measles mortality and morbidity, 1910-95. (A) Measles deaths by month, 1910-95. (B) *Main graph* Annual measles morbidity 1912-95 plotted as cases per 100,000 population. *Inset graph* Reported cases in thousands for the years 1980-95. Sources: Cliff, Haggett and Smallman-Raynor (1993, Figures 9.2 and 9.33, pp. 220, 224 and new data).

3.2 Autocorrelation on Graphs

The simplest model of the spread of disease from place to place in a country which reflects mixing between carriers of a disease (infectives) and those at risk (susceptibles) is a contagious diffusion process. Propagation occurs from the initial point of introduction to geographically-contiguous centres. These, in their turn, transmit the disease to their geographically nearest neighbours and so on.

Such a spatially highly-ordered process is, however, rarely found in practice. One of the commonest circumstances under which a contagious diffusion model is inappropriate is when transmission takes place through a system of cities. The varying population sizes of the settlements result in spread being directed from larger places to smaller. Typically, the initial point of introduction of a disease in a region is its largest city. Then urban centres next in size follow, and a 'drain down' process is initiated. In these circumstances, transmission is spatially discrete and fragmented rather than continuous because settlements at high levels of the population size hierarchy are generally some distance apart.

Diffusion processes often display both hierarchical and contagious elements. Simultaneous with the downwards percolation of disease from large settlements to small, spread also frequently takes place from each centre to its nearest neighbours.

3.2.1 Method

The method of autocorrelation on graphs is described in detail in Haggett (1976) and Cliff, Haggett, Ord and Versey (1981, pp. 99-102). The approach requires us to construct graphs consisting of nodes that represent the spatial structure to be tested, and lines or edges connecting the nodes. The connections are made so that the pattern of edges corresponds with the diffusion model we wish to test. Accordingly, to assess the importance of the diffusion models specified above in accounting for the spread of measles in the US, three transmission graphs were constructed in which the nodes represented the US states:

(1) A nearest neighbour graph defined by setting each element, w_{ij}, in a matrix $\mathbf{W}$ equal to 1 if states i and j shared a segment of common state boundary, and $w_{ij} = 0$ otherwise.

(2) A hierarchical graph in which $w_{ij} = 1$ if state j was the next larger or the next smaller state in population size to state i and $w_{ij} = 0$ otherwise; 1995 populations were used.

(3) Combined nearest neighbour/hierarchy graph in which $w_{ij} = 1$ if state j was the geographically nearest state that was either larger or smaller in population size than state i, and $w_{ij} = 0$ otherwise.

To determine the goodness-of-fit between each of these graphs and measles morbidity, the spatial autocorrelation coefficient, I, was computed for each month of the time series. Following Cliff and Ord (1981, pp. 17-21), we define I as

$$I = \left(\frac{n}{S_0}\right)\sum_{i=1}^{n}\sum_{j=1}^{n} w_{ij} z_i z_j \Big/ \sum_{i=1}^{n} z_i^2 \qquad (2)$$

where n is the number of states, x_i is the number or rate of measles cases per million population in state i in a given month of the time series, and $z_i = x_i - \bar{x}$. The $\{w_{ij}\}$ are drawn from graphs (1)–(3) above. In addition, $S_0 = \sum_{i=1}^{n}\sum_{j=1}^{n} w_{ij}$, $\bar{x} = \frac{1}{n}\sum_{i=1}^{n} x_i$, and we adopt the convention that $w_{ii} = 0$. I may be tested for significance as a standard Normal deviate. The expectation and variance under the null hypothesis of no spatial autocorrelation are given in Cliff and Ord (1981, p. 21). The greater the degree of correspondence between a given graph and measles morbidity, the larger will be the spatial autocorrelation coefficient.

3.2.2 Results

Fig. 5 shows the monthly standard Normal deviates for I for each of the three graphs, plotted as line traces against a backcloth of the reported monthly measles case rates, 1962-95, for the United States as a whole. Prior to plotting the Standard Normal deviates, the values were smoothed using a 5-point moving average. While this damps the amplitude of the curve (and so the apparent significance of individual values), the main cyclical events are more readily seen.

By treating the values for I on each graph as a time series, we can address a number of fundamental issues about the spread of measles epidemics. These questions are:

(1) Is each diffusion graph equally important over the study period, or are there time trends in relative importance?

(2) How does the monthly significance of I vary with changes in the monthly case rate?

(3) What are the epidemiological implications of these patterns?

We examine each question in turn.

Time trends

Inspection of Fig. 5 shows that, over the study period, there is a declining time trend for the nearest neighbour graph, a rising time trend for the population hierarchy graph, and a falling trend for the combined graph. These trends may be quantified by fitting the linear trend model,

$$SND(I_t) = b_0 + b_1 t + e_t \qquad (3)$$

where $SND(I_t)$ is the standard Normal deviate for I in month t, b_0 and b_1 are parameters to be estimated, and the $\{e_t\}$ are random disturbances.Fig. 6 plots the trend lines for each graph against a backcloth of reported case rates (stippled histograms). While linear trends characterised the time trend in autocorrelation for the nearest neighbour and combined nearest neighbour/hierarchy diffusion graphs, a quadratic time trend was required for the population hierarchy graph. The implication of the trends is that measles transmission over the study period shifted

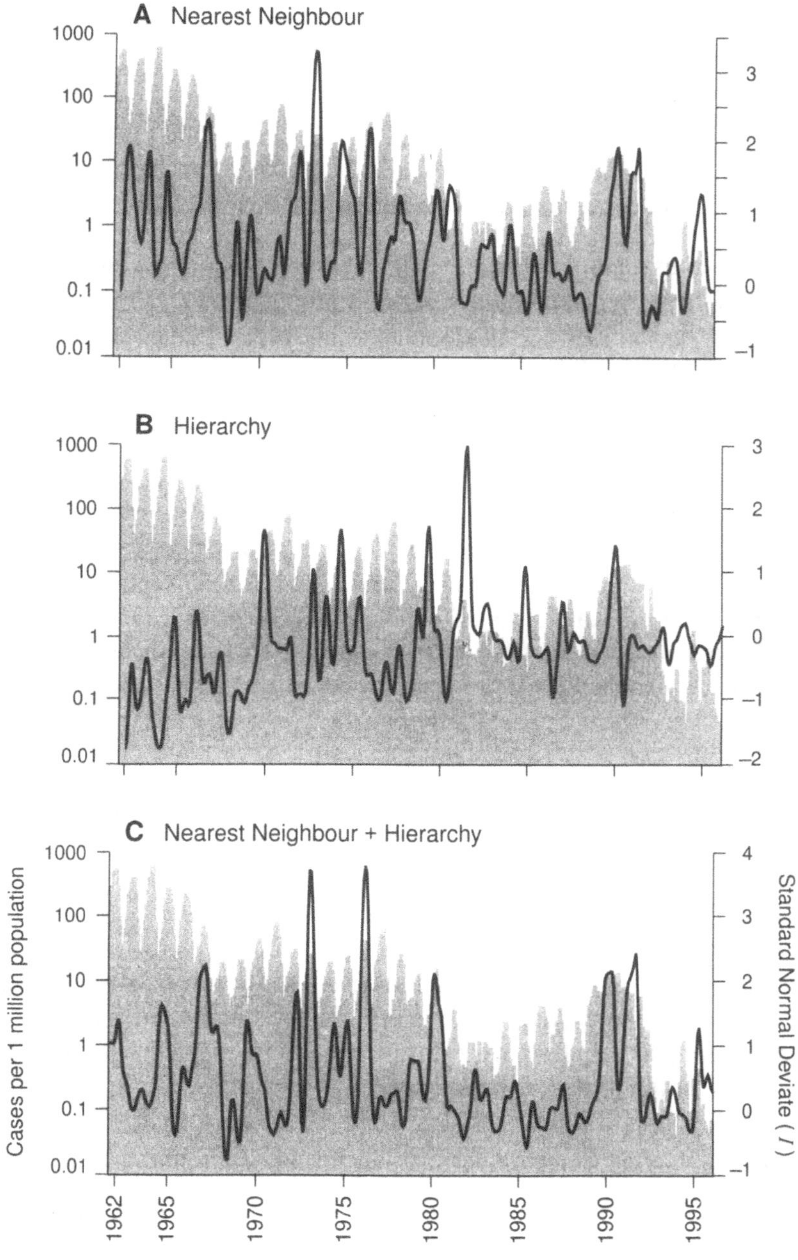

Fig. 5. Relationship between monthly reported cases of measles per million population in the United States, 1962-95, and levels of spatial autocorrelation. Histograms (stippled) give the case rates. Line traces plot the monthly values of the spatial autocorrelation coefficient, *I*, as standard Normal deviates for different diffusion graphs. (A) Nearest neighbour graph. (B) Population hierarchy graph. (C) Nearest larger/smaller state graph.

from one which was driven primarily by spatial transmission between states, to one in which drain down from larger to smaller states was also important. Reasons for this are discussed below.

Monthly variation

There are two aspects to this question: how does the relative importance of each graph vary over the course of the year?; and what are the phase relationships between monthly case rates and *I*? Monthly variations in *I* may be analyzed by calculating seasonal indices from the time series of *SND* values. The following steps were employed:

(a) After detrending as described above, the detrended data were smoothed by fitting a centred moving average of 12 (for monthly data).

(b) The moving average was subtracted from the detrended data to obtain the raw seasonals.

(c) For each month, the median value of the raw seasonals was calculated (i.e, the median of all the January values, 1962-95, and so on). These medians comprise the seasonal indices; see Chatfield (1980) or Box and Jenkins (1976) for details.

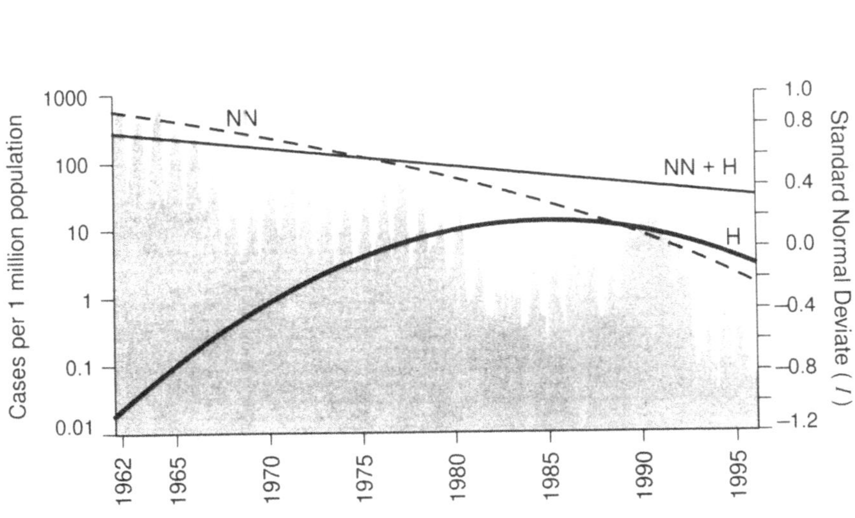

Fig. 6. Time trends in levels of spatial autocorrelation on different diffusion graphs for measles in the United States, 1962-95. Monthly case rates per million population are shown as stippled histograms. The trend lines for the spatial autocorrelation coefficient, *I*, expressed as a standard Normal score are plotted as line traces. NN=nearest neighbour graph. H=population hierarchy graph. NN+H=nearest larger/smaller state graph.

Fig. 7 plots the seasonal indices for each of the three diffusion graphs. It shows marked differences in the significance of each graph in different months of

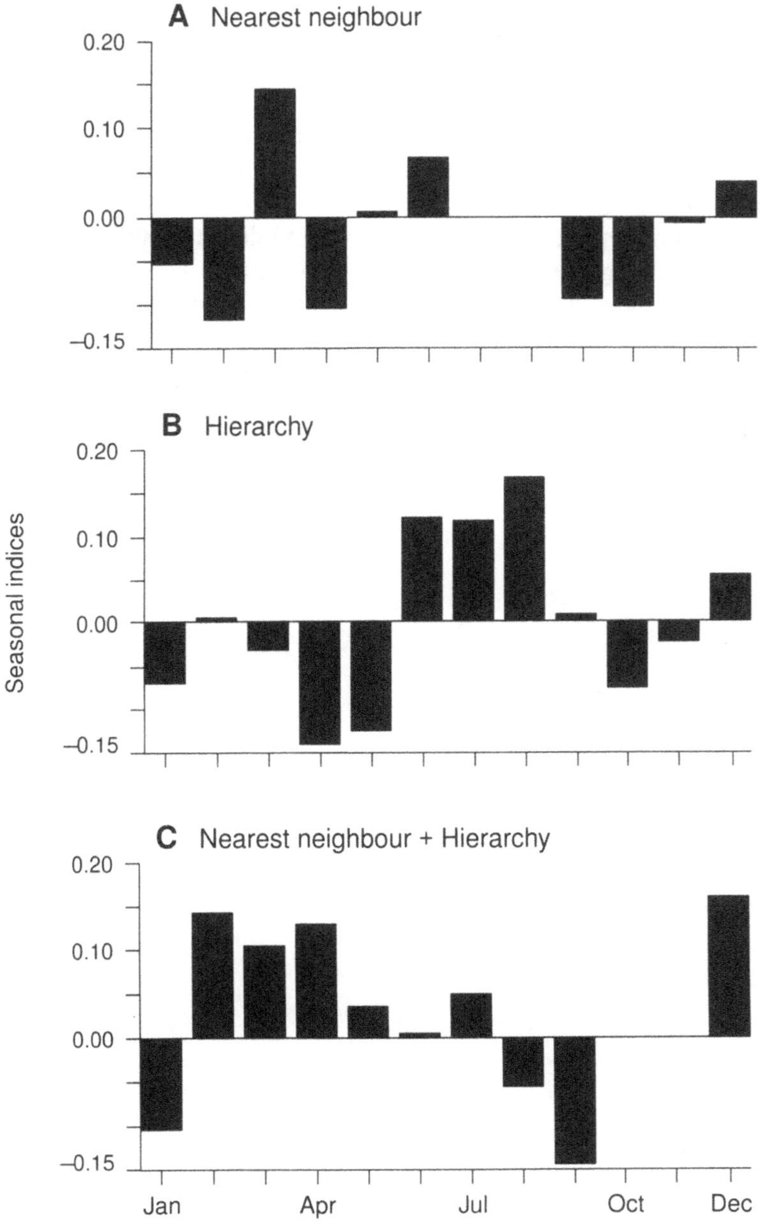

Fig. 7. Monthly excesses and deficits of spatial autocorrelation in reported measles cases per million population on different diffusion graphs, United States, 1962-95. (A) Nearest neighbour graph. (B) Population hierarchy graph. (C) Nearest larger/smaller state graph.

the year. During the 'normal' measles season in the United States of late winter/spring, a transmission process operating jointly through spatial spread and population size is clearly dominant (chart C). In the summer months of traditionally low measles case rates, propagation is dominated by population size (chart B). Significant purely spatial effects (chart A) seem to operate only at the switch over between winter/summer and summer/winter. Together, charts A–C imply that geographical transmission of measles between states occurs mainly in the main measles season and that, away from the spring peak of cases, the disease sinks back to persist in the principal population centres.

Phase relationships between the monthly time series for *SND(I)* and case rates may be established by calculating the cross correlation function (CCF) between the two. Cross-correlation analysis proceeds by computing the correlation coefficient, r_k, between any pair of time-series to determine the value of k at which the maximum correlation occurs. This value of k is conventionally taken as the lead or lag of the one series with respect to the other. If this occurs at $k = 0$, the two time series are said to be in phase.

Fig. 8 plots the CCFs between *SND(I)* and case rates for $-30 \le k \le 30$ for each of the diffusion graphs. It shows a strong relationship between case rates and the diffusion graphs containing some element of spatial transmission (charts A and C), and no phase relationships for the population size hierarchy graph. The implication is that, during measles epidemics, spatial transmission is dominant. If we examine the shapes of the CCFs with respect to the $k = 0$ position (epidemic peaks), then other interesting features are evident:

(i) On chart A, the nearest neighbour graph is significant over most of the months tested. The CCF is not symmetric; higher values are found in the months after epidemic peaks (lag 0) than before, implying that spatial effects are stronger as epidemics wane than they are in the build-up to peak case rates.

(ii) Chart C shows the same asymmetry, along with a cyclical element at 12 months and harmonics. This 'burst' of importance as part of the annual rhythm of measles epidemics is consistent with the seasonal importance of the same graph in Fig. 7C.

3.2.3 Epidemiological Significance

A full understanding of the epidemiological significance of Figures 5–8 is to be obtained from the spatial implications of work by Bartlett (1957, 1960), Black (1966) and Schenzle and Dietz (1987) on the critical community size for measles endemicity, and a study by Griffiths (1973) on the effect of vaccination rates upon this size threshold.

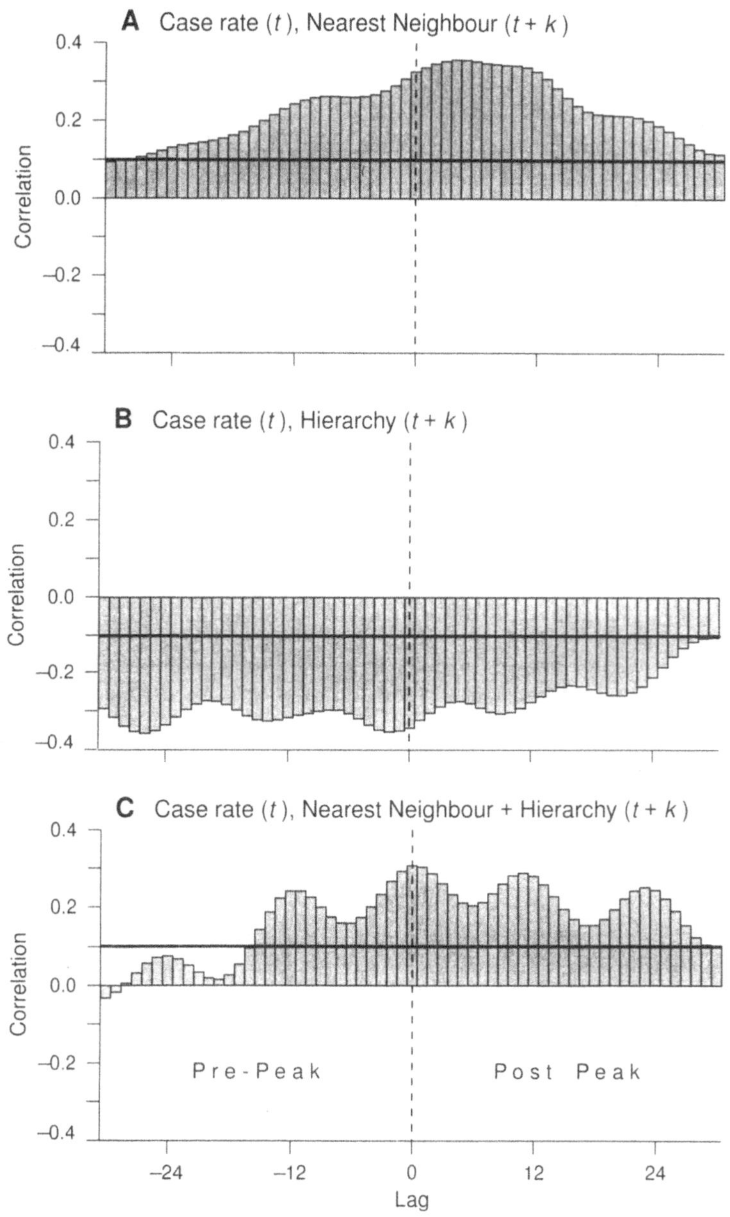

Fig. 8. Spread of measles in the United States. Cross-correlation functions between the monthly reported cases per million population and the standard Normal deviate for the spatial autocorrelation coefficient, *I*, calculated for different diffusion graphs. (A) Nearest neighbour graph. (B) Population hierarchy graph. (C) Nearest larger/smaller state graph.

Epidemic disease thresholds

Using a sample of 19 English towns and the virus disease of measles, Bartlett (1957) plotted the mean period between epidemics, in weeks, against the population size of these towns. Bartlett discovered that, in large towns, measles was endemic with periodic eruptions. In cities below a certain population size threshold, measles displayed an epidemic pattern with complete disappearance of the disease (fade-out) between epidemics. A distinction was drawn between urban areas above about 10,000 people with a regular pattern of epidemics, and those where occasional epidemics were missed, giving a more irregular pattern. Given the rates of reporting at the time of Bartlett's original study, the endemic/epidemic threshold for measles lay at a population size of around 250,000, a figure which has been confirmed on many occasions since (but which assumes no vaccination and no inter-area spread; see Cliff and Haggett (1990) for a recent review).

The existence of an endemicity threshold for a transmissible disease has profound implications for its persistence and control. In large geographical units above the size threshold, like community **A** in Fig. 9, a continuous trickle of cases is reported. These provide the reservoir of infection which sparks a major epidemic when the susceptible population, *S*, builds up by births and susceptible in-migrants to a critical level. When the *S* population reaches this level, an epidemic occurs upon the introduction of infection. The *S* population is diminished and the stock of infectives, *I*, increases as individuals are transferred by infection from the *S* to the *I* population. This generates the characteristic 'D'-shaped relationship over time between sizes of the *S* and *I* populations shown on the end plane of the block diagram.

When the total population of a community falls below the endemicity threshold, as in areas **B** and **C** of Fig. 9, measles epidemics recur only when the virus is reintroduced by the influx of infected individuals from reservoir areas (**A** in Fig. 9). These movements are shown by the broad arrows. In such smaller communities, the *S* population is insufficient to maintain a continuous record of infection. The disease dies out and the *S* population grows in the absence of infection. Eventually the *S* population will become big enough to sustain an epidemic when an index case arrives. It is the time–space repetition of these basic processes which generate the successive epidemic waves witnessed in most communities.

Nine years after Bartlett's original paper, his ideas were taken up in an island context by the Yale epidemiologist, Francis Black (Black, 1966). Black showed that the effect of geographical isolation was to increase the critical community size from c.250,000 to c.500,000.

Effect of vaccination

Griffiths (1973) has studied the impact of vaccination upon the critical community size. Suppose that x denotes the proportion unvaccinated in a population. Griffiths showed that the effect on the critical community size is to multiply it by a factor of $1/x^2$. For example, 90 percent vaccination (i.e., $x = 0.1$) would push the threshold up to 25 million.

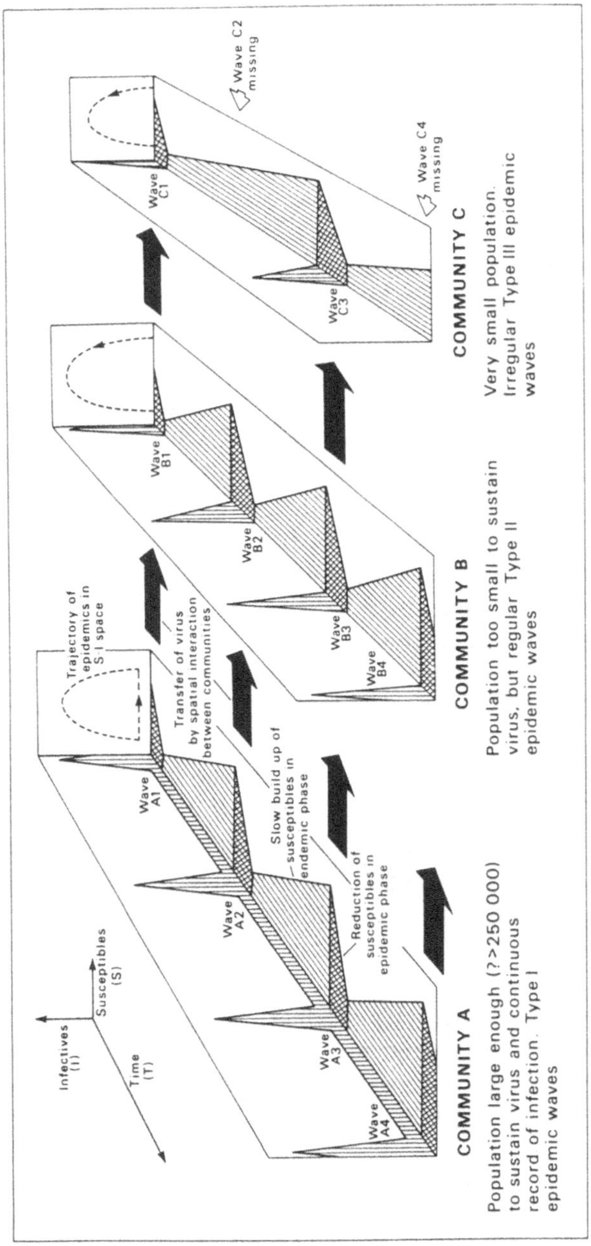

Fig. 9. Bartlett model of measles spread through communities of different population sizes. Source: Cliff and Haggett (1988, Fig. 6.5A, p. 246).

In the US context, by inspecting the measles time series for individual states, it is simple to identify the years in which no cases were being reported for the first time for several months in succession (taken as four in this paper). This gives some idea of the years in which vaccination had forced the critical community size up to the population total of each state. Fig. 10A plots these dates against $100(1-x)$, the percentage of the population needing to be vaccinated to push the critical community size up to the 1966 state populations. Circle sizes are proportional to the percentage of the total vaccination doses administered nationally between 1978 and 1988 in each state. Graph B plots the same data, but circle sizes have now been made proportional to the 1966 state populations. Chart A identifies the states which crossed the endemicity threshold late in the campaign and which remained the reservoir states (the type A communities of Fig. 9) for propagation of measles to those states which had eliminated endemic measles early in the campaign (type B communities in Fig. 9). The association of larger circles with the states that reached the threshold relatively late in the campaign indicates that the states needing the largest number of vaccinations to prevent endemicity were those with the most intensive vaccination programmes. The similar pattern of circle sizes in chart B implies these were also the states with the biggest populations. Thus Fig. 10 shows that measles was, from the mid 1970s, progressively pinned back into a limited number of reservoir states with large populations which acted as seeds for periodic national epidemics; these states were the subject of especially intensive vaccination programmes.

Discussion

These time-space patterns offer an explanation for the findings illustrated in Figures 5-8. High levels of vaccination over the last three decades have destroyed the simple inter-state propagation of the disease, and placed increasing emphasis upon the demographic reservoir areas as the disease has been pinned back into fewer and fewer geographical locations. This is evidenced by the upwards trend in the significance of the population hierarchy graph in Fig. 6, and the parallel downwards trend in the significance of the nearest neighbour graph since vaccination began. Fig. 7 also highlights the importance of the population reservoirs for maintaining measles in the United States in inter-epidemic periods. The population hierarchy graph is significant during the summer months of each year, and spatial propagation switches in during epidemics to produce dispersion from the reservoir areas, a direct parallel of the mechanisms illustrated in Fig. 9. The switching on of spatial transmission during epidemics is confirmed by the strong correlations between the nearest neighbour graph and the waxing and waning of case rates illustrated in Fig. 8A. The lack of association between changes in case rates and the hierarchy graph of Fig. 8B is consistent with the continuous presence of measles in the endemic reservoirs.

3.3 Centroid Trajectories

We can gain further insights into the relative importance of contagious and hierarchical diffusion effects at various stages in the US measles elimination campaign by computing the mean centre of measles activity in each month of the time series, and monitoring the temporal oscillations of this mean. Kuhn and Kuenne (1962) give an appropriate centroid statistic. Let the location of the *j*th

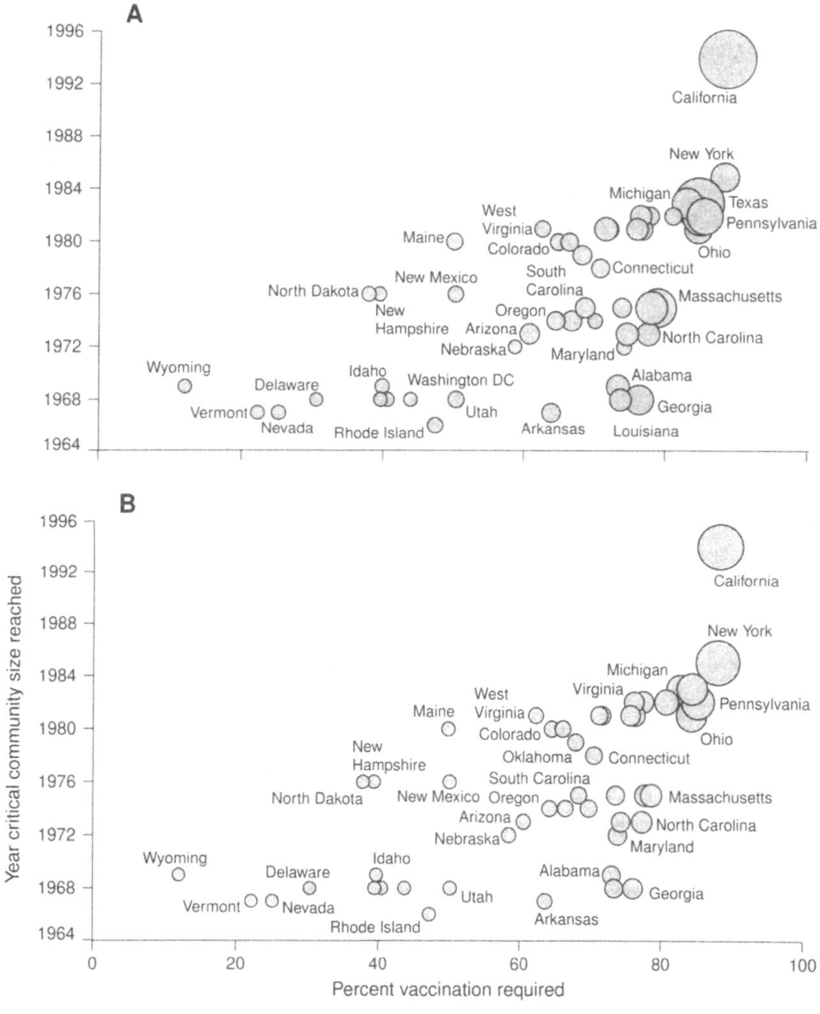

Fig. 10. Critical community size and measles vaccination programmes in the United States, 1962-95. (A) Year in which states first conquered endemic measles (vertical axis) against percentage of the state population needing vaccination to achieve this (horizontal axis). Circles sizes are proportional to percent of national public-sector vaccination doses administered, 1978-88, in each state. (B) As (A), but circle sizes are proportional to state populations.

state be given a horizontal cartesian co-ordinate u_j and a vertical co-ordinate v_j (say for its geographical centre). Let the reported number of cases (per unit of population) for the *j*th state be I_j. Choose an initial location for the centroid at a map co-ordinate position $\hat{U}, \hat{V}$. The centroid of the distribution is now located by an iterative search procedure based upon repeated solution of the equations

$$\hat{U}_{k+1} = \sum_{j=1}^{n} I_j u_j d_{j(k)} / \sum_{j=1}^{n} I_j d_{j(k)} \text{, and } \hat{V}_{k+1} = \sum_{j=1}^{n} I_j v_j d_{j(k)} / \sum_{j=1}^{n} I_j d_{j(k)} . \quad (4)$$

until $\hat{U}_{(k+1)} - \hat{U}_{(k)} < \varepsilon$ and $\hat{V}_{(k+1)} - \hat{V}_{(k)} < \varepsilon$ where ε is a pre-specified convergence error level. In these equations, d_j is the distance between the latest centroid and the *j*th area. The new centroid is denoted by the subscript $(k + 1)$. By plotting centroids for successive periods and linking them in sequence, the general direction of movement can be captured. While the location of the geographical centre of an outbreak at any point in time obviously captures only one limited aspect of the spread of disease, it does tell us where the action is concentrated at any particular moment in time.

Using reported cases per million population, equation (4) was calculated for each of the 408 months in the US measles time series. Fig. 11 shows the latitude and longitude of the disease centroid for two representative months of each year, 1962-95. February has been taken to illustrate the behaviour of the centroid in the measles epidemic season, and July to show its behaviour during the summer months away from epidemic episodes. The graphs have been oriented to correspond to latitude and longitude. To aid interpretation, the latitudes and longitudes of representative states have also been marked.

Fig. 11 shows a fundamental shift in the centroid behaviour from the second half of the 1970s. Prior to this date, the centroid in both February and July generally occurred in the Midwest. In terms of longitude, after 1975, it oscillated between the densely populated states of the Midwest/Northeast and the West; these oscillations began somewhat earlier for the February than for the July centroid. For latitude, similar oscillations began between the heavily populated states of the north and south of the country. This switch in behaviour is consistent with the increasing significance of hierarchical as opposed to contagious diffusion found with the autocorrelation on graphs approach; recall that Figures 5 and 6 showed population effects increased rapidly up to 1980 as mass vaccination programmes reduced the number of states with endemic measles. We may interpret Fig. 11 as showing, after 1975, the way in which the maintenance of measles within the US depended upon the large population reservoirs, flipping between these centres once nation-wide endemic measles had been conquered.

4 Area-based Data. II: Non-linear Geometries for Measles in the United States, 1962–95

The analyses described have illustrated some of the complex variations that have taken place in the space-time behaviour of measles epidemics in the United States over the last 30 years. One way of modelling what is evidently non-linear behaviour is to use multidimensional scaling (MDS). A full account of the methodology involved appears in Torgerson (1958) and Coxon (1982). In this section, we outline the main features of the method, and apply the technique to the US measles data.

MDS refers to a family of statistical methods by which the information

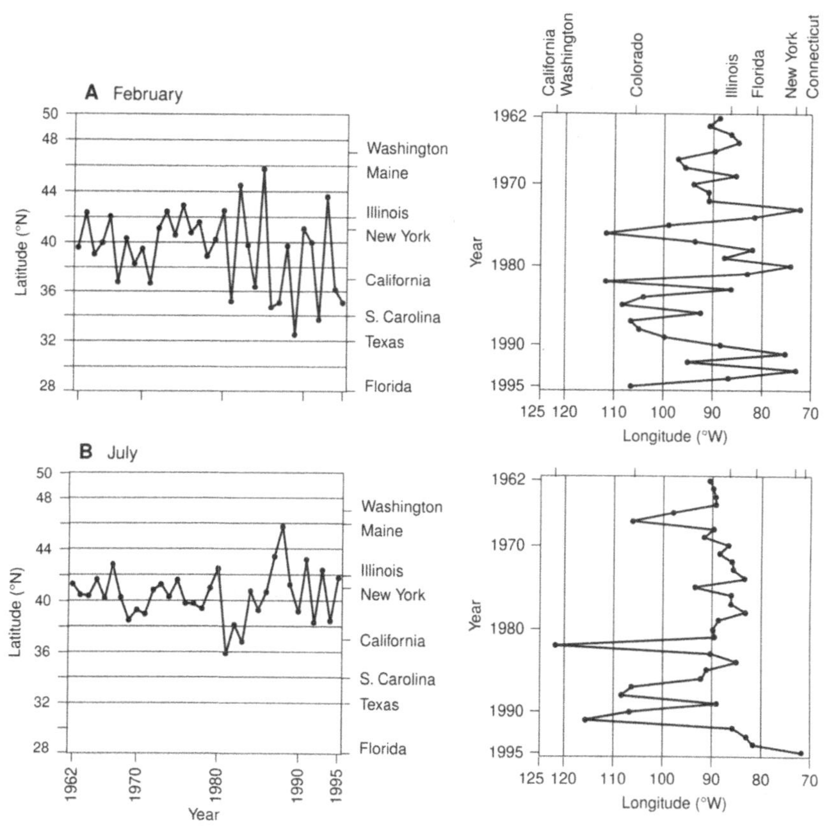

Fig. 11. Kuhn-Kuenne centroids. Time changes, 1962-95, in the February (A) and July (B) centroids of measles activity in the United States. Left hand diagrams give the time trajectory of the latitude of the centroid; right hand diagrams give the longitude. Based upon monthly reported cases per million population.

contained in a set of data is represented by a set of points in multidimensional space. These points are located in the space in such a way that the geometrical relationships between them reflect empirical relationships in the data. The term, *multidimensional*, denotes the fact that the space into which the points are fitted may vary from low order spaces (for example, one-dimensional scales or two-dimensional 'maps') up to higher order spaces (for example, three-dimensional 'cubes' or four- or more dimensional hyperspaces). See Kruskal and Wish (1978) and Coxon (1982) for reviews of the mathematics of MDS.

4.1 The MDS Mapping Concept

The essential idea behind MDS is most readily illustrated using a geographical analogy. When a conventional map of a portion of the earth's surface is drawn, some distortion results in transforming a curved segment of the globe onto a flat piece of paper. The particular map projection used will determine the nature and extent of the scale or directional distortion introduced. But, subject to these known errors, all the map projections in common use attempt to ensure that the locations of points on the map systematically reflect their relative positions on the globe. If, however, we use MDS to represent points on the globe, a map is constructed in which the locations of the points on the map correspond not to their (scaled) geographical locations but to their degree of *similarity* on some variable measured for them. For example, we might map points into a time space; locations separated by short travel times will appear central and close together in such a space, while centres linked by long travel times will be widely separated and peripheral.

This basic idea can be extended to the mapping of points from geographical space into a 'disease space'. Points with, say, similar attack rates for an infectious disease will be located close to each other on the MDS map even though they may be far removed geographically. The greater the degree of similarity between places on the variable measured, the closer together the places will be in the MDS space. Conversely points which are dissimilar on the variable will be widely separated in the MDS space, irrespective of their geographical location on the globe. Thus, broadly stated, the problem of MDS is to find a configuration of n points in m-dimensional space such that the interpoint distances in the configuration in some sense match the experimental (dis)similarities of the n objects. For geographical problems, MDS mapping in two dimensional spaces (m = 2) is especially valuable. This will yield an MDS map upon which the relative locations of points, defined in terms of similarity of their time series of reported case rates of a disease, may be compared with their locations on a conventional geographical map. Indeed, by starting each MDS run with the areas in their natural geographical positions, it is possible to track the progressive adjustment from the geographical map to a best-fit epidemiological space.

The goodness-of-fit between the MDS configuration and the observed (dis)similarities is measured via *stress* (low stress reflects good fit on a scale from 0 to 1). Each configuration is normalized by placing its centre of gravity or centroid at the origin and then stretching or shrinking so that the mean squared distance of the points from the origin equals 1.

4.2 The US Measles Data Revisited

Our analysis of these data so far has led us to suggest that the progressive elimination of endemic measles in the American states by vaccination led, from c.1980, to increasing epidemiological significance for a few very large states that persisted as measles reservoirs. These states acted as diffusion poles from which spatial transmission occurred by contagious and hierarchical effects during general epidemic episodes. We may test this qualitative model using MDS as follows:

(i) The time series was divided into two halves, (a) 1962-80, during which most states experienced endemic measles for much of the time, and (b) 1981-95 when endemic measles had been eliminated from all states except the very large ones such New York and California.

(ii) For each half, the coordinates of the states in 1- and 2-dimensional MDS spaces were calculated. The Pearson product moment correlation coefficient, r_{ij}, computed between all $\{i,j\}$ pairs of US state time series of monthly reported measles cases per million population, was used as our measure of similarity. Random starting coordinates were used to initiate the solution procedure for the 1-dimensional fit; the latitudes and longitudes of the state centres were used to start the 2-dimensional fit.

(iii) To examine the relationship between state positions in the MDS space and population size, we focused upon the 1-dimensional fit and calculated the Pearson correlation coefficient between state coordinates on the line and their populations [in 1980 for period (a) and 1995 for period (b)]. If population size effects grew in importance between periods (a) and (b), we would expect the correlation to increase.

(iv) To examine the persistence of spatial effects between periods (a) and (b), we used a procedure described in Tober (1965) to compute the degree of spatial correlation between the latitude/longitude coordinates (x,y) and the final (u,v) coordinates in the MDS space. The closer the MDS coordinates are to the true latitude/longitudes, the higher will be the spatial correlation (and the more 'map-like' the MDS configuration).

Spatial goodness-of-fit

Tobler's method is based upon the affine transformation. The beauty of the affine transformation in its full form is that, given two sets of coordinates, the transformation matrix handles all aspects of the tranformation (scaling, shear, rotation, translation, projection and overall scaling). Thus it is possible to compare the relationship between the same set of objects mapped in two different coordinate systems.

Let $x_i, y_i; u_i, v_i$ denote the rectangular coordinates of the *i*th point in the two point patterns we wish to compare (here the longitudes and latitudes of the centroid of the *i*-th state and the corresponding coordinates in the MDS space, respectively). In addition, let $W_i = (u_i, v_i)'$ and $Z_i = (x_i, y_i)'$ represent the *i*th observation pair, $i = 1, 2, \cdots, n$, where the primes denote the transpose. The objective is to use least squares methods to estimate the coefficients A and B in the transformation

$$\hat{W}_i = A + BZ_i \,. \tag{5}$$

The regression can be considered as a mapping from the xy plane to the uv plane.

Treating the observations as the components of vectors in this way means the constants become

$$A = (a_1, a_2)' \text{ and } B = \begin{pmatrix} b_{11} & b_{12} \\ b_{21} & b_{22} \end{pmatrix}, \tag{6}$$

and the equation to be minimized is

$$\sum_{i=1}^{n} \left(\hat{W}_i - W_i\right) \cdot \left(\hat{W}_i - W\right)' . \tag{7}$$

The complete affine transformation is therefore of the form

$$\begin{pmatrix} \hat{u}_j \\ \hat{v}_j \end{pmatrix} = \begin{pmatrix} a_1 \\ a_2 \end{pmatrix} + \begin{pmatrix} b_{11} & b_{12} \\ b_{21} & b_{22} \end{pmatrix} \begin{pmatrix} x_i \\ y_i \end{pmatrix}. \tag{8}$$

Defined in terms of simple correlations, the affine correlation, which is an overall measure of the degree of correlation between the two point patterns, is given by

$$R_{WZ} = \frac{\sigma_u^2\left(r_{xu}^2 + r_{yu}^2 - 2r_{xu}r_{yu}r_{xy}\right) + \sigma_v^2\left(r_{xv}^2 + r_{yv}^2 - 2r_{xv}r_{yv}r_{xy}\right)}{\left(\sigma_u^2 + \sigma_v^2\right)\left(1 - r_{xy}^2\right)} . \tag{9}$$

This is not symmetric and so $R_{WZ} \neq R_{ZW}$.

Table 1 and Figs. 12 and 13 give the results obtained. The diagrams show the configurations of the states in the various MDS spaces, while Table 1 gives the results of the correlation analyses. For the 1-dimensional spaces, Fig. 12 shows somewhat greater separation and population size ordering of the states along the line in the second half of the study period as compared with the first. This is confirmed in Table 1 where the correlations between state locations and population size are both low, but double as between the first period and the second. This strengthening of population size effects over time is consistent with the results of the tests for spatial autocorrelation on the hierarchy graph illustrated in Fig. 6B.

Table 1 Propagation of measles in the United States, 1962-95. Correlations between MDS coordinates of states and (a) state populations (1-dimension), (b) latitudes and longitudes of state centres (2-dimensions).

	MDS solution(1-dimension)	MDS solution (2-dimensions)	
Time period	r(coordinates, populations)	R_{WZ}	R_{ZW}
1962-80	0.13	0.11	0.23
1981-95	0.25	0.33	0.31

Table values give correlations between state coordinates in the minimum stress MDS spaces and (a) state populations for the 1-dimensional solution, and (b) the latitude and longitudes of the geographical centres of states for the 2-dimensional solution. $n = 49$.

For the 2-dimensional spaces (Fig. 13), there are distinct differences between the MDS configurations. In the earlier period (chart A), there is a large cluster of states around the origins of the space. The states in this group are primarily the heavily populated states of the East and West coasts and the Midwest. Around this hub is rim of dispersed states drawn geographically from the central United States from the Mississippi valley to the Rockies. Thus the groupings in the MDS space seem to be by population size. In the later period from 1981-95 (chart B), the range of coordinates is less and states are more evenly distributed through the space; the central hub and rim of chart A has disappeared.

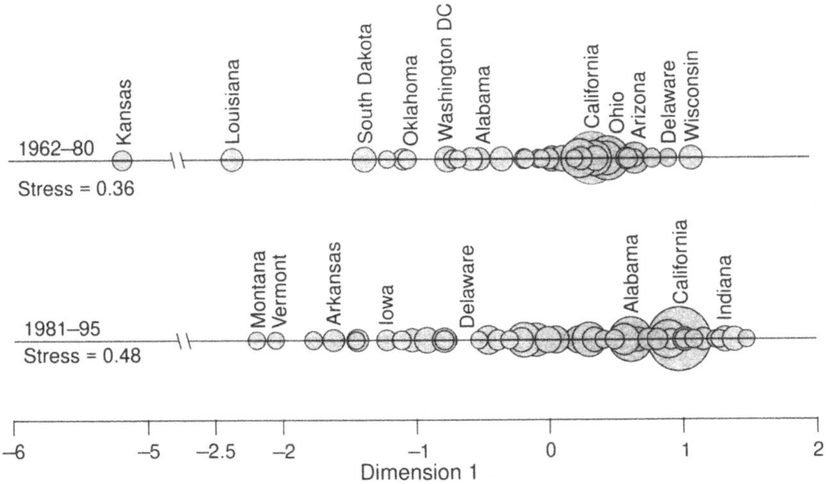

Fig. 12. One-dimensional minimum-stress MDS configurations for US measles data (case rates per million population). (A) 1962-80. Circles are proportional to state populations in 1980. (B) 1981-95. Circles are proportional to population sizes in 1995.

The disintegration of the central hub between the two time periods implies that the state time series have become less similar and operate more independently of each other in the second period that in the first. This is consistent with the more map-like configuration (higher affine correlation in Table 1) in the second period – states show a greater tendency to function as independent geographical units. Nothwithstanding these changes, most of the states with large populations remained concentrated around the origin of the MDS space.

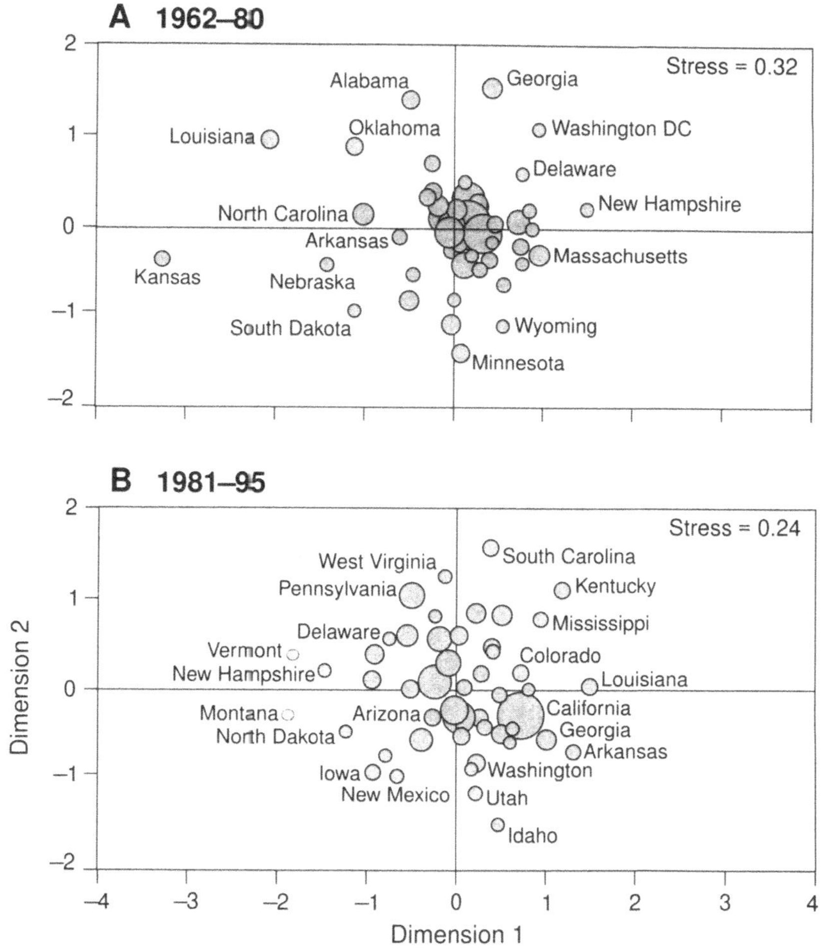

Fig. 13. Two-dimensional minimum-stress MDS configurations for US measles data (case rates per million population). (A) 1962-80. Circles are proportional to state populations in 1980. (B) 1981-95. Circles are proportional to population sizes in 1995. The initial configuration used to start the solution algorithm consisted of the longitudes and latitudes of state centres.

5 Area-based Data. III: External Corridors for Measles in the United States, 1962–95

By the mid-1980s, indigenous case levels of measles in the United States had been so reduced by vaccination that attention turned, in the elimination programme, to external sources of infection. Such imports will be related, obviously, both to the level of measles incidence in the foreign country and to the level of exchange in terms of travel flow between that country and the United States. The specific risk in a particular part of the United States can be linked to the concentration of flow from a foreign country into particular gateway cities and states where, for example, major international airports are located.

Markowitz, Tomasi, Hawkins, *et al.*, (1988) have studied international measles importations into the United States between 1980 and 1985. In the five years, three countries accounted for over one third: Mexico, the UK and the Philippines. Over half of the total importations were recorded in just six states: California, New York, Maryland, Florida, Texas and Hawaii. If these findings on international importations are linked with those of Sections 3-5 which show the way in which measles has been maintained and propagated internally in the United States, then we can summarise the spatial processes involved in Fig. 14.

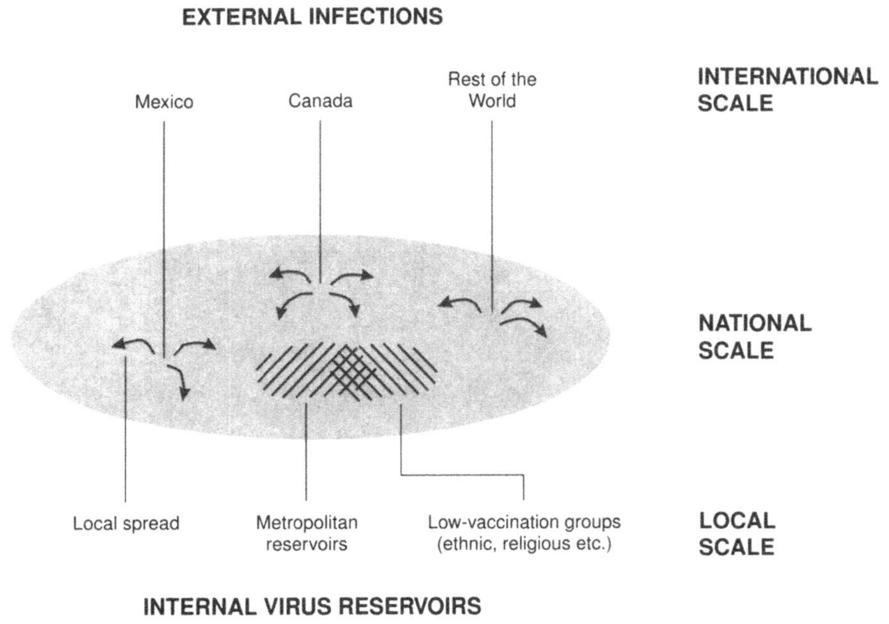

Fig. 14. Schematic diagram of the spatial processes influencing the distribution of measles within the United States. Source: Cliff and Haggett (1995, Fig. 4, p. 418).

Despite the massive reductions in measles morbidity in the United States over the last thirty years through vaccination, the virus may still have the opportunity to persist internally in pockets in the large and densely populated states, renewed by the continuing influx of measles from other countries. Reductions in external virus invasions will depend on WHO's Expanded Programme of Immunization.

6 Conclusion

This paper has illustrated the some of the ways in which the patterns in geocoded point and area based data may be analyzed. Taking John Snow's classic data set showing the geographical locations of cholera deaths in the Soho district of London in 1854 as an example of point pattern data, it has been shown how very simple mapping techniques point unerring to the Broad Street pump indicted by Snow as the seat of the epidemic. The clustering of deaths around this source was then confirmed using point pattern methods.

The area-based data used in this paper gave, for the United States, the monthly reported cases of measles per million population by state for the period, 1962-95. Using a variety of spatial analytical methods–autocorrelation on graphs, disease centroids, and multidimensional scaling–it was show how the propagation of measles epidemics in the United States became increasingly dependent upon spatial transmission from relatively few endemic reservoirs located in the largest states. The significance of these reservoirs grew over the study period as the vaccination campaigns designed to eliminate indigenous measles began to achieve their objective after 1980. A second consequence of this campaign was to enhance the importance of measles transmission into the United States from other countries in which measles remains a significant disease.

Modern disease surveillance and mapping is increasingly dependent upon the use of Geographical Information Systems (GIS). The display capabilities of GIS for showing disease patterns against covers of co-factor variables are self-evident. But much remains to be done before visual spatial associations can be readily tested for epidemiological significance within a GIS, and issues of real as opposed to spurious correlations fully addressed. A variety of methods should always be used to cross-check the spatial patterns believed to exist in disease data. For it is only by such careful cross-checking that we can avoid generating the apparently significant mapped distributions that not only send researchers off on wasteful wild goose chases but also bring the geographical approach to the analysis of disease data into disrepute.

References

Bartlett, M.S. (1957). 'Measles periodicity and community size.' *Journal of the Royal Statistical Society A*, **120**, 48-70.

Bartlett, M.S. (1960). 'The critical community size for measles in the United States.' *Journal of the Royal Statistical Society A*, **123**, 37-44.

Black, F.L. (1966). 'Measles endemicity in insular populations: critical community size and its implications.' *Journal of Theoretical Biology*, **11**, 207-11.

Box, G.E.P., Jenkins, G.M. (1970). *Time Series Analysis, Forecasting and Control.* San Francisco: Holden Day.

Chatfield, C. (1980). *The Analysis of Time Series: An Introduction* (2nd edition). London: Chapman and Hall.

Cliff, A.D, Haggett, P. (1988). *Atlas of Disease Distributions: Analytic Approaches to Epidemiological data.* Oxford: Blackwell Reference.

Cliff, A.D., Haggett, P. (1990). 'Epidemic control and critical community size: spatial aspects of eliminating communicable diseases in human populations'. In *London Papers in Regional Science* (ed. R.W. Thomas), **21**, London: Pion, pp. 93-110.

Cliff, A.D., Haggett, P., Ord, J.K., Versey, G.R. (1981). *Spatial Diffusion: an Historical Geography of Epidemics in an Island Community.* Cambridge: Cambridge University Press.

Cliff, A.D., Haggett, P., Smallman-Raynor, M.R. (1993). *Measles: an Historical Geography of a Major Human Viral Disease from Global Expansion to Local Retreat, 1840-1990.* Oxford: Blackwell Reference.

Cliff, A.D., Ord, J.K. (1981). *Spatial Processes.* London: Pion.

Coxon, A.P.M. (1982). *The User's Guide to Multidimensional Scaling with Special Reference to the MDS(X) Library of Computer Programs.* London: Heinemann.

Diggle, P.J. (1983). *Statistical Analysis of Spatial Point Patterns.* New York: Academic Press.

Griffiths, D.A. (1973). 'The effect of measles vaccination on the incidence of measles in the community.' *Journal of the Royal Statistical Society A*, **136**, 441-9.

Haggett, P. (1976). 'Hybridizing alternative models of an epidemic diffusion process.' *Economic Geography*, **52**, 136-46.

Kruskal, J.B., Wish, M. (1978). *Multidimensional Scaling.* Beverly Hills, Ca.: Sage Publications.

Kuhn, H.W., Kuenne, R.E. (1962). 'An efficient algorithm for the efficient solution of the generalized Weber problem in spatial economics.' *Journal of Regional Science*, **4**, 21-33.

Markowitz, L.E., Tomasi, A., Hawkins, C.E., Preblud, S.R. Orenstein, W.A., Hinman, A.R. (1988). 'International measles importations United States, 1980-1985.' *International Journal of Epidemiology*, **17**, 187-92.

Ripley, B.D. (1976). 'The second-order analysis of stationary point processes.' *Journal of Applied Probability*, **13**, 255-66.

Schenzle, D., Dietz, K. (1987). 'Critical population sizes for endemic virus transmission'. In *Räumliche Persistenz und Diffusion von Krankheiten* (eds. W. Fricke & E. Hinz), Heidelberg: Heidelberg Geographical Studies, **83**, 31-42.

Senser, D.J., Dull, H.B., Langmuir, A.D. (1967). 'Epidemiological basis for the eradication of measles.' *Public Health Reports*, **82**, 253-6.

Snow, J. (1849, 1854). *On the Mode of Communication of Cholera* (1st & 2nd editions). London: Churchill Livingstone.

Tobler, W.R. (1965). 'Computation of the correspondence of geographical patterns.' *Regional Science Association, Papers*, **15**, 131-42.

Torgerson, W.S. (1958). *Theory and Methods of Scaling.* New York: Wiley.

A Spatio-Temporal Model: An Application to the AIDS Epidemic in São Paulo, Brazil

Célia Landmann Szwarcwald, DrPH and Francisco Inácio Bastos, DrPH
Department of Health Information, Fundação Oswaldo Cruz.
Av. Brasil, 4365 RJ 21045-900 Rio de Janeiro, Brazil
e-mail:celia@malaria.procc.fiocruz.br

Abstract: *In this paper, a spatio-temporal model is proposed in order to distinguish spatial diffusion patterns of diseases. The procedure is based on the definition of different connection matrices that allow us to test for temporal and space-time interaction effects on disease rates. When applied to the AIDS epidemic in São Paulo, Brazil, strong differences emerge regarding distinct categories of exposure. While the AIDS epidemic among IDUs follows an "in situ" growth model, the results for the heterosexual transmission group strongly suggest a spatial diffusion process. From the epidemiological point of view, these findings indicate that preventive strategies should be urgently renewed.*

1. Introduction

Spatial analysis is an attractive methodology for research areas that deal with geographically referenced observations. Important applications include epidemiological investigation. In the course of time, geographical variation of mortality and incidence rates of certain diseases have proven to be valuable clues to disease causation[8]. In addition, the recent development of statistical techniques that incorporate dependence in space has reinforced its use in disease studies[7].

Since data on disease are often recorded in a set of geographical units during several time periods, examining the pattern of the disease rates by considering not only dependence in space but also dependence in time and space-time interdependence may be of great interest. In particular, estimation of temporal, spatial and interaction effects on incidence rates may contribute to our understanding of the spreading process of the disease.

In this paper, a method of analysis of geographically referenced data recorded in two successive time periods is proposed. It allows us to test for temporal and space-time autocorrelation of disease rates. Data on the AIDS epidemic in São Paulo, Brazil are used to illustrate the procedure.

2. Previous Studies

To address the question of whether spatial-time clustering is present in a series of occurrences, Knox[4] (1964) proposed a space-time interaction coefficient based on spatial and temporal distances between occurrences. Knox's coefficient was generalized by Mantel[6] (1967) by using the reciprocal transformations of the distances as measures of spatial and temporal closeness. Since then, for point pattern data, several other procedures have been proposed to test for the presence of spatial-time clustering[9,10,12].

Extending earlier work by Cliff and Ord[1] (1973) for lattice data, Hagget[3] (1976) examined the spatial diffusion pattern of measles in Cornwall, 1969-70 using a binary autocorrelation statistic. The binary coding recorded each subarea as infected or vacant during a 222-week period.

A generalized space-time correlation measure was also used to study the diffusion pattern of the cholera epidemic in London, 1849[2]. Using information on weekly death rate from cholera in each of the London boroughs, the binary coding expressed the occurrence of high death rates (above the mean rate) in contiguous areas recorded at two successive weeks. The importance of "in situ" growth and spatial diffusion was further studied through the application of an ordinary least-square stepwise regression procedure.

In a recent work, using spatial correlograms computed from yearly incidence rates by county, Lam et al[5] (1996) studied the spatio-temporal spread of the AIDS epidemic in four regions of the United States from 1982 to 1990.

In the present study, based on Cliff and Ord[2] (1981) methodology, we propose a spatial-temporal model for lattice data. The procedure is based on the definition of different connection matrices that quantify temporal and space-time effects. The model may contribute to the understanding of disease diffusion patterns when applied to disease rates. We use the technique to study the AIDS epidemic spread in São Paulo municipalities, from 1988 to 1995, concerning distinct exposure categories.

3. Preliminaries

Consider the linear model

$$(1)\quad \boldsymbol{y} = m\,\boldsymbol{1} + \boldsymbol{\epsilon}$$

where $\boldsymbol{y}$ is the response vector of n observations and $E(\epsilon) = \boldsymbol{0}$.

A well-known statistic that may be applied on the model residuals to test for the existence of spatial correlation is the generalized Moran's I coefficient[2].

It is defined as:

$$(2)\quad I=\frac{n\sum_{i=1}^{n}\sum_{j=1,j\neq i}^{n} w_{ij}e_ie_j}{w_{..}\sum_{i}^{n} e_i^2},\quad for\ e_i=y_i-\bar{y},\ \bar{y}=\frac{1}{n}\sum_{i=1}^{n} y_i,\ w_{..}=\sum_i\sum_j w_{ij},$$

where w_{ij} is some measure of spatial proximity.

Using matrix formulation, the vector of residuals may be written as:

$$(3)\quad \boldsymbol{e}=\boldsymbol{My},\ where\ \boldsymbol{M}=\boldsymbol{I}-\frac{1}{n}\boldsymbol{11'},\ \boldsymbol{I}=identity\ matrix,\ \boldsymbol{1'}=(1,\ldots,1).$$

The statistic I is of the form:

$$(4)\quad I=\frac{\boldsymbol{e'We}}{\boldsymbol{e'e}},\ for\ \boldsymbol{W}=(w_{ij}),\ the\ matrix\ of\ spatial\ weights,$$

and the weights are scaled so that $w_{..}=n$.

Under the assumption that ϵ has a multivariate normal distribution $N(\boldsymbol{0}, \sigma^2\boldsymbol{I})$, it can be shown that I is asymptotically normally distributed under some fairly mild conditions[2]. An approximate test of significance is therefore provided by evaluating the I statistic as a standard normal deviate. Also, since the generalized Moran's coefficient is a scale-free function of independent normal variates, the Pitman-Koopmans theorem gives that I is distributed independently of $Q=\boldsymbol{e'e}$, so that $E(I^p)=E((\boldsymbol{e'We})^p)/E((\boldsymbol{e'e})^p)$. But from Searl[11] (1971), the rth cumulant of $\boldsymbol{e'Ae}$ is:

$$(5)\quad K_r(A)=2^{r-1}(r-1)!\,tr(\boldsymbol{AV})^r,\ for\ \boldsymbol{V}=\sigma^2\boldsymbol{M}.$$

This gives:

$$(6)\quad E(I)=\frac{tr(\boldsymbol{WV})}{tr(\boldsymbol{V})}\quad and$$

$$(7)\quad E(I^2)=\frac{2tr(\boldsymbol{WVWV})+(tr(\boldsymbol{WV}))^2}{2tr(\boldsymbol{V})+(tr(\boldsymbol{V}))^2},$$

which provides the mean and variance of the coefficient I.

4. A space-time model

We shall now consider the case of observations recorded in terms of their location in two successive periods of time. We denote the vector of observations as:

$$(8) \quad y' = (y_{11} \dots y_{n1}, y_{12} \dots y_{n2}) = (y_1', y_2')$$

where y_j is the vector of observations at time j, j=1,2.

Model (1) is now expressed as:

$$(9) \quad y = \begin{bmatrix} 1 & 0 \\ 0 & 1 \end{bmatrix} \begin{bmatrix} m_1 \\ m_2 \end{bmatrix} + \epsilon, \ \text{for } 1 = \begin{bmatrix} 1 \\ \cdot \\ \cdot \\ 1 \end{bmatrix} \ \text{and } 0 = \begin{bmatrix} 0 \\ \cdot \\ \cdot \\ 0 \end{bmatrix}.$$

Then, the vector of residuals is equal to:

$$(10) \quad e = \begin{bmatrix} e_1 \\ e_2 \end{bmatrix} = \begin{bmatrix} M & O \\ O & M \end{bmatrix} \begin{bmatrix} y_1 \\ y_2 \end{bmatrix}, \ \text{for } M = I - \frac{1}{n} 1'1 \ \text{and } O = \begin{bmatrix} 0 & . & . & 0 \\ . & . & . & . \\ 0 & . & . & 0 \end{bmatrix}.$$

In order to examine the problem by considering the effect of time alone and the space-time interaction effect, we propose the use of two different connection matrices. To define the temporal correlation or the correlation between observations taken over two successive time intervals, we consider as the connection matrix:

$$(11) \quad W = \begin{bmatrix} O & I \\ I & O \end{bmatrix}, \ \text{for } I = \text{identity matrix}.$$

The observation in area i at t=2 is connected to observation in the same area i at t=1 and the temporal correlation is equal to:

$$(12) \quad r_t = \frac{2\, e_1' e_2}{e_1' e_1 + e_2' e_2}.$$

For this particular case:

(13) $$V=\sigma^2\begin{bmatrix} M & O \\ O & M \end{bmatrix},\quad WV=\sigma^2\begin{bmatrix} O & M \\ M & O \end{bmatrix},\quad WVWV=\sigma^4\begin{bmatrix} M & O \\ O & M \end{bmatrix},$$

(14) $$tr(WV)=0,\quad tr(WVWV)=2\sigma^4(n-1),$$

(15) $$tr(V)=2\sigma^2(n-1),\quad tr(V^2)=2\sigma^4(n-1).$$

So,

(16) $$E(r_t)=0 \quad and \quad E(r_t^2)=\frac{1}{n}.$$

To search for the space-time effect, we consider the following connection matrix:

(17) $$W=\begin{bmatrix} O & W_s \\ W_s & O \end{bmatrix},\quad for\ W_s=matrix\ of\ spatial\ weights,$$

where the observation in area i at t=2 is joined to observations in neighbor areas j at t=1.

The space-time index is written as:

(18) $$r_{st}=\frac{2\,e_1{}'W_s e_2}{e_1{}'e_1+e_2{}'e_2}.$$

For this particular case,

(19) $$V=\sigma^2\begin{bmatrix} M & O \\ O & M \end{bmatrix},\quad WV=\sigma^2\begin{bmatrix} O & W_sM \\ W_sM & O \end{bmatrix},\quad WVWV=\sigma^4\begin{bmatrix} W_sM & O \\ O & W_sM \end{bmatrix},$$

$$(20)\quad tr(WV)=0,\quad tr(WVWV)=2\sigma^4 tr(W_s M W_s M),$$

$$(21)\quad tr(V)=2\sigma^2(n-1),\qquad tr(V^2)=2\sigma^4(n-1).$$

Therefore,

$$(22)\quad E(r_{st})=0,\quad and\quad E(r_{st}^{\ 2})=\frac{tr(W_s M W_s M)}{n^2-1}\ .$$

The index r_t describes the importance of the "in situ" growth while $r_{s\text{-}t}$ describes the strength of the spatial diffusion. Since each one of those autocorrelation indexes reduces to the coefficient given in (4), the corresponding normal approximations will apply.

Now, if we want to determine if the process is influenced by both a local growth and a spatial diffusion process, we propose the use of a connection matrix that incorporates both effects, that is:

$$(23)\quad W(b)=\begin{bmatrix} O & (1-b)I+bW_s \\ (1-b)I+bW_s & O \end{bmatrix},\quad for\ some\ b,\ 0\le b\le 1,$$

and W_s is a matrix of spatial weights where $w_{i.}=1$. So, the closer the value of b is to 1, the stronger is the space-time effect and vice-versa. The corresponding correlation coefficient is given by:

$$(24)\quad r_{t,st}(b)=(1-b)\,r_t+b\,r_{st}.$$

Since we want to find the maximum autocorrelation measure, we define $\hat{b}$ to be the value that maximizes the standard $r_{t,st}(b)$.

5.Application

5.1. Background

As of December, 1996 approximately 100,000 accumulated cases of AIDS were reported by the Brazilian National Coordination on STD/AIDS. Until 1987, the AIDS epidemic in Brazil was basically restricted to the two main Brazilian cities, São Paulo and Rio de Janeiro. However, from 1990 on, the AIDS epidemic has spread to all Brazilian regions.

Despite the significant spread of the epidemic throughout the country, the State of São Paulo alone has reported 53% of all Brazilian cases. The State of São Paulo remains the main focus of the epidemic not only because of the high incidence in the metropolitan area but also because of the spread of the disease throughout the mid-size municipalities. From 1990 to 1996, the number of small municipalities (50,000 to 200,000 inhabitants) with at least one AIDS case increased enormously.

The main purpose of this application is to show how the proposed space-time model may contribute to the understanding of the diffusion pattern of the AIDS epidemics in Brazil. The reason the state of Sao Paulo was selected for this application lies on the fact that it has the highest accumulated AIDS incidence rate and has one of the best notification systems in the country.

5.2. Material and Methods

In this study, we considered all reported AIDS cases in the state of Sao Paulo, Brazil from 1987 to 1995. The year of 1996 was not included in the analysis because of possible notification delay.

Information on AIDS (National Division of STD/AIDS-Brazilian Ministry of Health) and population data (Brazilian Geographic and Statistics Institute) were merged and made compatible. The merged database contains information on the frequency of AIDS cases classified by municipality, year of official notification and exposure category as well as basic demographic data and geographic coordinates.

For the sake of analysis, data were aggregated into two time periods: 1987-91 and 1992-95 and the square-root transformation was used as the variance-controlling transformation. This transformation was selected, as opposed to the logarithm transformation, because of the presence of many zero-incidence rates in the first period of time. So, as the response variable, we considered the square-root of the incidence rates calculated by exposure category and period of time for all municipalities with at least one AIDS case in the second period of time. For the definition of the spatial connection matrices, we used "first-nearest-neighbors" and binary weights ($w_{ij}=1$ if municipality j is the first-nearest neighbor of i, $w_{ij}=0$, otherwise).

5.3. Results

Thematic maps of the AIDS incidence rate by selected exposure categories are shown in Figure 1. Different spatial patterns are found. Among injecting drug users (IDUs), we observe that the phenomenon is concentrated in the metropolitan area and along the North, crossing several mid-size cities,

Figure 1: AIDS incidence rates (per 100,000) by selected exposure categories and periods of time. São Paulo, Brazil.

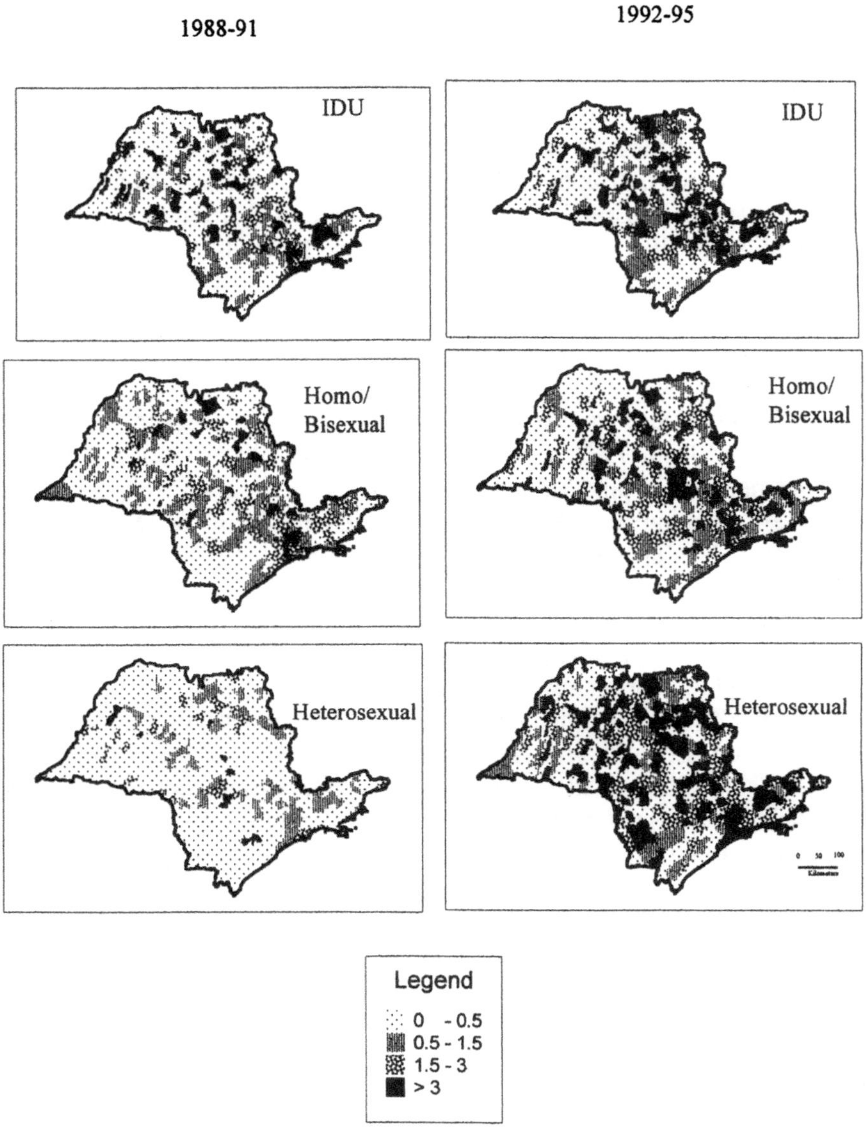

most with have a much higher socioeconomic level when compared to the rest of the country. On the other hand, the corresponding spatial configuration for the heterosexual transmission fails to show any predominant region. It has spread overall, mainly among the small municipalities, with less than 200,000 inhabitants.

Estimation of the proposed autocorrelation coefficients helps us to distinguish the spatial diffusion patterns of the different exposure categories (Table 1). For the total AIDS group, both time and space-time effects are highly significant, although the value of $\hat{b}$ (0.30) indicates a clear predominance of an "in situ" growth process. For the homo/bisexual category, no significant autocorrelation was found. In the group of IDUs, the epidemic is dominated by a localized growth with the lowest value of $\hat{b}$ (0.18). However, spatial proximity seems to have a relevant influence on the heterosexual epidemic spread. The space-time autocorrelation is significantly different from zero and the value of $\hat{b}$, close to 0.5, suggests that spatial diffusion is as important as "in situ" growth.

Table 1: Autocorrelation Indexes by Category of Exposure. AIDS Epidemic, São Paulo, Brazil, 1987-91 and 1992-95

Category of Exposure	Average N-N Distance	r_t	r_{st}	$\hat{b}$
Homo/bisexual (n=229)	16.1 km	0.068 (NS)	0.054 (NS)	0.49
Heterosexual (n=275)	15.2 km	0.202 (p=0.01)	0.142 (p=0.05)	0.46
IDU (n=226)	18.0 km	0.460 (p=0.00)	0.086 (NS)	0.18
TOTAL (n=410)	14.3 km	0.580 (p=0.00)	0.211 (p=0.00)	0.30

From the epidemiological point of view, these findings are specially interesting because they indicate that the geographic expansion of the AIDS epidemic is mainly due to heterosexual transmission. They highlight the fact that the epidemic is "leaking" from traditional exposure categories, shown to be more circumscribed in this analysis, toward general population.

6. Final remarks

In this paper, benefiting from the flexibility of the Cliff and Ord[2] methodology, we showed the possibility of checking for temporal and space-time interaction effects on disease rates. The procedure is based on the definition of a vector that incorporates observations in two successive time intervals and on the definition of some appropriate connection matrices.

Applying the proposed procedure to AIDS accumulated incidence rates recorded in São Paulo, Brazil, in two successive periods of time, 1988-91 and 1992-95, we could distinguish different patterns of spatial diffusion by category of exposure. The results suggested that in the geographic spread of the disease the heterosexual transmission prevails. While IDU cases appeared to be described by "in situ" growth processes, the group of heterosexual AIDS cases were marked by space-time effects.

The geographic expansion of the heterosexual transmission is worthy of public health concern, specially considering the low educational level of the Brazilian population and the high prevalence of other sexually transmitted diseases. Moreover, taking into account the social, cultural and economical subordination of the Brazilian women to their sexual partners, preventive efforts should urgently be rethought.

In summary, this application is a clear example of how a spatio-temporal model can contribute to the understanding of epidemic diffusion patterns and, consequently, to the implementing of more efficient programs to control the spread of diseases. Particularly in the case of AIDS, understanding the epidemic in its space-time dimension, specifically in developing countries, seems to be a promising field for investigation and further study.

References

1. Cliff A D & Ord J K, 1973. *Spatial Autocorrelation*. London, Pion.
2. Cliff A D & Ord J K, 1981. *Spatial Processes: Models & Applications*. London, Pion.
3. Hagget P, 1976. Hybridizing Alternative Models of an Epidemic Diffusion Process, *Economic Geography*, 52:136-146.
4. Knox E G, 1964. The Detection of Space-Time Interaction, *Applied Statistics*, 13: 25-29.
5. Lam N S-N, Fan M & Liu K-B, 1996. Spatial-Temporal Spread of the AIDS Epidemic, 1982-1990: A Correlogram Analysis of Four Regions of the United States, *Geographical Analysis*, 28: 93-107.
6. Mantel N, 1967. The Detection of Disease Clustering and a Generalized Regression Approach, *Cancer Research*, 27:209-220.
7. Marshall R J, 1991. A Review of Methods for the Statistical Analysis of Spatial Patterns of Disease, *Journal of the Royal Statistics Society, series A*, 154:421-441.
8. Mayer J D, 1983. The Role of Spatial Analysis and Geographic Data in the Detection of Disease Causation, *Social Science and Medicine*, 17: 1213-1221.
9. McAuliffe T L & Afifi A A, 1984. Comparison of a nearest neighbor and other approaches to the detection of space-time clustering, *Computational Statistics & Data Analysis*, 2: 125-142.
10. Raubertas R F, 1988. Spatial and Temporal Analysis of Disease Occurrence for Detection of Clustering, *Biometrics*, 44: 1121-1129.
11. Searle S R, 1971. *Linear Models*. New York, John Wiley & Sons.
12. Williams G W, 1984. Time-space clustering of disease. In: Cornell R G (ed), *Statistical Methods for Cancer Studies*, New york, Marcel Dekker: 167-227.

A Geographic Information System Approach to Mapping AIDS in England and Wales: Standardisation of a Heterogeneous Data Set

A M Molesworth*†, J F Raper†, D J Unwin†, B G Evans*

* Public Health Laboratory Service AIDS and STD Centre, London UK NW9 5EQ
† Department of Geography, Birkbeck College, University of London UK W1P 2LL

Abstract: Approximately 13 000 cases of AIDS were diagnosed and reported in adults in England and Wales to the end of June 1997. It was possible to consider the geographic distribution of cases at a higher resolution than currently used, by creating a dot plot of cases for which a National Grid reference to the Postcode Unit of residence at diagnosis was available. This method excluded 30% of the cases. Alternative methods need to be used which allow consideration of the problems of poor spatial resolution, poor geo-referencing and under-reporting of AIDS cases, at the same time assuring the maintenance of patient confidentiality.

Introduction. This poster introduces a research project to explore the geographic distribution of Human Immunodeficiency Virus (HIV)-related ill-health in England and Wales in order to enhance our understanding of the epidemic and its surveillance and to improve the quality of information relevant to Public Health specialists. It documents the preliminary process of assigning geographic location to case reports and discusses problems in mapping the disease distribution, at this stage based on the 13 000 cases of Acquired Immune Deficiency Syndrome (AIDS) reported in adults in England and Wales to the end of June 1997[1].

Background. Since 1982 surveillance of AIDS in England and Wales has been necessary to monitor the epidemic, to inform care and prevention initiatives and to project the future burden of infection and ill-health. It involves the confidential and voluntary reporting, by clinicians, of newly diagnosed cases to the Public Health Laboratory Service (PHLS) AIDS and STD Centre. Information routinely available to Public Health specialists considers geographic variation in disease distribution, but the need to ensure patient confidentiality restricts this information to counts of cases within large administrative health areas. Any spatial patterns observed may thus be more a product of the definition of administrative areas than of the actual underlying distribution of disease. Patient-based information is further limited because its circulation is restricted beyond those directly responsible for patient care. Many patients attend centres outside the health area where they live and so

Public Health specialists may have inadequate information on how the epidemic is affecting their resident population [2]. Using Postcodes it may be possible to consider the geographic distribution of AIDS cases at a greater resolution than is currently used.

Methods. Reports of AIDS cases include a reference to where they were reported as living at the time of diagnosis. This may refer to one or more of the areal-based spatial units described below (table).

Table: Levels of the areal-based spatial unit stratification used to describe case locations ***(Source: Royal Mail Address Manager 1997/1)***

Level of areal aggregation	Units in England &
NHS Executive Region	10
Health Authority	100
Local Authority (includes Unitary Authorities)	350
Postcode District	2 500
Postcode Units (residential)	1 500 000

† Approximations (April 1997)

The geographic location of residence reported for each case was referenced to one of these spatial units. For over 9000 (70%) cases the Postcode Unit allowed a case to be assigned to the other, lower resolution, spatial units by reference to the Postzon File supplied by the Royal Mail. The same process allowed a National Grid reference to be assigned to each Postcode Unit. For the remaining 4000, the location of residence could not be defined by the Postcode Unit, or the National Grid reference. This proportion was lower for cases reported recently (figure).

The disease distribution was then visualised in ArcView 3.0 as a dot plot of those cases for whom a point reference on the National Grid was available (not shown).

Discussion. The need to maintain patient anonymity restricted the presentation of information using dot plot maps, since every dot represented a reference to where a patient was reported living at AIDS diagnosis. There were further problems associated with the allocation of cases to point references. The point coordinate assigned to a Postcode Unit (Postcode) represented the National Grid reference of the south west corner of the 100m grid square in which the first address is reported to lie. For many Postcodes the grid-referencing is known to be worse than implied by the 100m resolution of the data, and it is biased towards the first address in the postal delivery round [3]. It is not known to what extent this may have affected the accuracy of mapping disease distribution. Multiple cases with the same Postcode, and at 100m resolution more than one Postcode, may share the same National Grid reference and have been represented by a single point reference in the dot plot. Alternative methods need to be considered which overcome this problem, as well as that of assuring patient confidentiality. It was further noted that any spatial patterns observed in the data failed to include diagnosed but, as yet, unreported cases and would therefore tend to under-estimate the spatial case distribution. In addition, of those cases reported 30% had no point reference and could not be presented in the dot plot, this proportion varying also with time. Future investigation of the spatial case distribution needs to account for the spatial and temporal heterogeneity of the data, and consider the spatial distribution of disease rates and its implications for resource allocation.

Figure: Proportion of areal-based spatial units reported for location of residence by unit and year of report

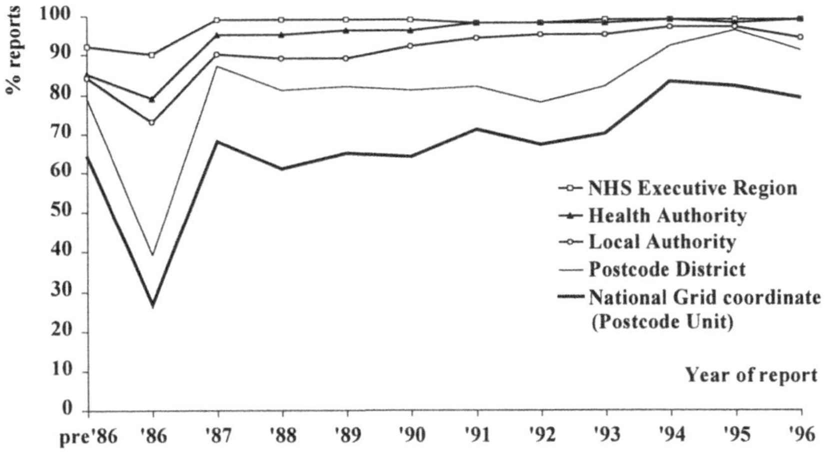

Conclusion. The method allowed high resolution visualisation of the case distribution independent of the administrative areas previously used, providing a truer reflection of disease distribution. However its failure to include all diagnosed AIDS cases, or those with non-point-based spatial references, and to adjust for known inaccuracies in assigning location may have introduced unknown bias into any observed spatial case distributions, and at this stage demands a cautious interpretation of results. It was suggested that a kernel estimation approach to future visualisation and analysis could overcome or minimise some of these problems. Although there are problems associated specifically with this methodology, such an approach would enable analysis through high resolution mapping at the same time maintaining confidentiality.

References:

[1] PHLS AIDS and STD Centre - Communicable Disease Surveillance Centre. *Unpublished Quarterly Surveillance Tables No.36, June 1997*

[2] Molesworth A. Results of the 1995 survey of prevalent clinically diagnosed HIV infection in England, Wales, and Northern Ireland. *Commun Dis Rep CDR Rev 1997; 6: R77-82*

[3] Raper J, Rhind D, Shepherd J. Postcodes: The New Geography *Longman Scientific and Technical 1992*

Exploratory Spatial Data Analysis of Motor Neurone Disease in North West England: Beyond the Address at Diagnosis

Clive E. Sabel [1] and Anthony C. Gatrell [2]

Department of Geography [1]
Institute for Health Research [2]

University of Lancaster
Lancaster
LA1 4YB
England.

c.sabel@lancaster.ac.uk
a.gatrell@lancaster.ac.uk

This paper addresses the issues surrounding an individual's exposure to potential environmental risk factors which can be implicated in the aetiology of a disease. We hope to further elucidate the 'lag' or latency period between the initial exposure to potential pathogens and the physical emergence of the disease, with specific reference to the rare neurological condition, Motor Neurone Disease (MND). A Space-Time approach is adopted, whereby patterns in both time and space are considered. No prior assumptions about the aetiology of MND are adopted. By using methods for the analysis of point processes, which preserve the continuous nature of the data, we resolve some of the problems of analysis based on often arbitrary areal units, such as postcode boundaries, or political boundaries, which diseases do not respect.

We use kernel estimation to model Space-Time patterns. Raised relative risk is assessed by adopting appropriate adjustments for the underlying population at risk, with the use of controls. Significance of the results are assessed using Monte-Carlo simulation.

Our results demonstrate the utility of kernel estimation as a visualisation tool. Small areas of elevated risk appear to have been identified, which need to be more closely examined before any firm conclusions can be drawn. We highlight a number of issues concerning the inadequacies of the data, and possibly of the techniques themselves. Various modifications have been proposed, with the most promising being adaptive kernel estimation.

Keywords: Space-Time clustering, Kernel Estimation, GIS, Motor Neurone Disease.

1. Introduction

Geographical epidemiology rests largely upon the assumption that the spatial incidence of diseases holds a key to their causes (Schæström, 1996). However, high mobility, long latent periods and environmental change complicate matters, distorting what might otherwise be a direct relationship between cause and effect. This gives rise to what Schæström has called Space-Time lag. From a geographical point of view, this means that the place or environment where the case is discovered and diagnosed is not necessarily the same place or environment where the exposure occurred (Picheral, 1982).

Many studies examining associations between geographical patterns of disease and causal factors adopt the key underlying assumption that current residence in an area can be equated with exposure to conditions that currently (and historically) pertain there (Bentham, 1988). This is important, since the place of residence at the time of diagnosis or death is often adopted by epidemiologists and geographers as the location for further analysis of the disease in question. Yet people move, and hence previous exposure to pathogens will not be included in the study. The problems will be greater for diseases that have a long lag or latency period, allowing plenty of time for mobility of the population. By adopting only the current residential address, not only will an individual's migration history be neglected, but additionally the daily 'activity spaces' of the patient will be ignored.

Existing studies of disease clustering are often criticised due to the way in which boundaries of space and time are chosen. Some commentators have even questioned the value of investigating individual disease clusters at all, due to methodological shortcomings (Rothman, 1990). This study hopes to avoid these traditional criticisms, by adopting an individual or point approach, with no spatial aggregation. It is precisely this which Schæström called for when he designed a methodology that is 'as released as possible from preclusions in the shape of artificial spatial and temporal units and population aggregates' (Schæström, 1996).

Specifically, this research intends to search for any clustering which may suggest infective mechanisms at specific ages or dates in addition to spatial patterning. Hence the adoption of the Space-Time approach as advocated by Löytönen (1996) and Schæström (1996). It is hoped that this approach will provide a better insight into possible aetiological factors.

In order to counter criticisms such as those highlighted by Bentham (1988) and Rothman (1990), and to address the issues discussed above, regarding exposure and lag period definition, it is essential to work on a longitudinal data set which includes complete residential histories of patients, not just current residences. By taking time or age slices (residences at a given date or age of patients) and then mapping these results using Geographical Information Systems (GIS), insight into lag periods and exposure to potential pathogens might be gleaned. The disease under investigation here is Motor Neurone Disease (MND).

2. Methods

Our underlying philosophy is to maintain the data in its original individual form. This non-aggregated approach, using methods for the analysis of point processes, which preserve the continuous nature of the data, resolves some of the problems of analysis based on often arbitrary areal units such as postcode boundaries, or political boundaries. As such we hope to avoid the problems which often beset area based approaches and can lead to the Modifiable Areal Unit Problem (MAUP), whereby possibly false interpretations are made from analyses purely as a result of arbitrary aggregations of the data.

2.1 Georeferencing

In the UK, (and increasingly elsewhere), it is now possible to extract a full postcode from any given address, and match this postcode to the UK National Grid coordinate system, which geocodes each postcode to a spatial accuracy of at least 100m (Gatrell, 1989). Each unit postcode represents on average 15 households (slightly more in urban areas, where population is more concentrated, and less in rural areas where the population is more dispersed). It is these coordinates, together with length of residence, which are used here.

2.2 Testing Space-Time interaction

As Bithell (1990) has highlighted, the classic 'Knox' method for testing space-time interaction without knowledge of the population distribution is inappropriate for non-contagious diseases. Bithell proposed adopting density estimation techniques, which provide a well-founded approach to the problem of estimating relative risks at different geographical locations. Bithell's ideas have been developed further by Kelsall and Diggle (1995) and here we extend them to deal with the temporal component on the third dimension. This is achieved by treating the data items not just as individual cases, but by attaching to each of these locations a value, representing the length of residence at that location. This procedure essentially weights the input items as an input to the density estimation. The null hypothesis which we will therefore be testing is that of constant risk.

2.3 Kernel Estimation

Kernel estimation is a statistical technique whereby in epidemiological applications a distribution of discrete points or 'events' representing incidence of disease is transformed into a continuous surface of disease risk. Essentially, a moving three-dimensional function (the kernel) of a given radius or 'bandwidth' 'visits' each of the points or events in turn, and weights the area surrounding the point proportionately to its distance to the event. The sum of these individual kernels is then calculated for the study region, and a smoothed surface produced. There are a variety of different kernels. The one used by ARC/INFO, and adopted for our study, is the bivariate biweight kernel. This kernel has the property of being computationally simple, and is hence attractive for implementation within a GIS where large data sets are not uncommon. It can be defined as:

$$K(x) = \frac{3}{\pi}(1-x^t x)^2$$

where x represents the multivariate data set whose underlying density we are estimating. It is generally accepted that the choice of kernel function is not critical (Silverman, 1986; Kelsall and Diggle, 1995). However, the choice of bandwidth is crucial. The bandwidth has been variously referred to as the window width, smoothing parameter or radius (as referenced in ARC/INFO's Pointdensity function). Too large a bandwidth will over-smooth the data, giving very broad trends only, whereas under-smoothing will maintain localised variation but render interpretation more difficult. The problem of choosing how much to smooth is of crucial importance in density estimation (Silverman, 1986). Often a subjective judgment is made; indeed Silverman suggests that this might even be desirable where the purpose of the exercise is to explore the data in order to suggest possible hypotheses, as is the case in this study.

If the geographical distribution of the underlying population at risk was uniform, then a simple kernel of the patients would suffice. However, population is not uniformly distributed, and so we must consider methods for estimating the background risk. One method of getting around this problem could be to transform the point-pattern of disease by the underlying population distribution, i.e. by creating a cartogram based on population. Any patterning then observed would be immediately obvious and real. This is an area for further research.

2.5 Ratio of kernels

The problem now becomes one of estimating the population at risk and making suitable corrections, which we have attempted to do by using a sample of controls. *A priori*, we would expect disease intensity to vary with the underlying population density (i.e. to expect 'clusters' in population centres) but how can we test if the observed risks are significantly different from what was expected? By taking a ratio of the kernel estimates of case and control intensity, we can produce a map of relative risk and identify 'peaks' and 'troughs' in the data. It is worth noting that identifying 'gaps' in relative disease risk might be just as illuminating in developing the aetiology of the disease as examining the 'clusters'. In our particular case the third dimension, representing height, will be the variable of analysis of the kernel, and represents the length or period of residence of an individual at a location. Hence longer periods of residence at a location will be used to weight the analysis; these will have greater influence upon the kernel surface than short residential periods.

When taking the ratio of kernels, the issue of bandwidth has an added significance. We are now concerned here with the ratio of two kernel estimates of intensity. It does not necessarily follow that 'good' estimates of the numerator and denominator will lead to a good estimate of their ratio. For instance, small changes in the denominator (the estimate of background intensity) in regions where its value is small, will produce dramatic and unacceptable variations in the ratio. Bailey and Gatrell (1995) suggest that it might indeed be preferable to deliberately 'over-smooth' the kernel estimate of the background intensity when estimating the ratio, by adopting a larger bandwidth than would be normally appropriate if one were just interested in an estimate of the population intensity alone.

These ideas were initially exploited by Bithell (1990) in a study of clustering of Childhood Leukaemia. As an exploratory device, the approach has much to commend itself, although as Bithell points out, controls must be selected with care and the choice of bandwidth is critical to the outcome.

2.6 Significance Testing

If we want to interpret the results from a ratio of kernels, that is, our estimate of relative risk, it is important to have some idea of the statistical significance of local peaks and troughs under the null hypothesis of constant risk. As we have already noted, we would expect to observe clustering due to the natural background variation in the population from which events arise. What we are more interested in however, is detecting evidence of an elevated risk over and above this underlying environmental heterogeneity.

Bailey and Gatrell (1995) suggest treating the cases and controls from a study region as belonging to the same sample distribution. If there is no clustering of 'cases' relative to 'controls', this amounts to 'cases' being just a random sample from the pattern of both the cases and controls. Thus the hypothesis requiring testing is that of random 'labelling' of the cases and controls. By adopting the methodology of Kelsall and Diggle (1995) for each case-control pairing, we randomly assign one of the pair to be a simulated 'case', and the other a simulated 'control'. If we then repeat this for all pairings, we obtain a simulated case and a simulated control map. We can then generate the relative risk maps as before. Using this random allocation idea, we then repeat the process in a form of Monte-Carlo simulation. After performing, say, 999 iterations, we can establish upper and lower simulation envelopes; we then have an estimate of how unusual the observed pattern is. If the observed pattern lies outside the simulation envelope, one can begin to speak of areas of significantly elevated, or reduced, risk..

The results of the simulations can be graphically displayed by constructing a '*p*-value surface' which gives the proportion of simulated cells which are less than each observed cell, for each grid cell in the matrix. The 2.5% and 97.5% contours of this surface can then be draped over the original map of estimated risk, to highlight regions which correspond to significantly high or low risk. Kelsall and Diggle (1995) refer to these contours as 95% pointwise tolerance contours.

2.7 Software environments

The raster GRID component of ESRI's ARC/INFO GIS was used for all kernel estimation calculations. This was achieved using the Pointdensity function, taking the kernel option. This functionality is available in ARC/INFO version 7.1.1. ESRI's ArcView 3.0a was also used for display manipulation and hardcopy output.

3. Application to environmental epidemiology

Having established some of the fundamental concepts and methods which we have adopted and developed, we now consider the application of these ideas, using data on the incidence of MND in Lancashire and south Cumbria, England.

3.1 Motor Neurone Disease

Motor Neurone Disease (MND), also known as Amyotrophic Lateral Sclerosis (ALS) is a rapidly progressive, unremissive, fatal neuromuscular disease, with no known cause or cure. The absence of any real knowledge of aetiology makes it particularly attractive for the exploratory analysis of the form we are proposing.

Reported annual incidence of MND varies from 0.07 per 100,000 population in Mexico to 2.6 per 100,000 in Sweden. Patients are primarily aged 50-75, with maximum incidence at 65 years of age (Foster, 1992). Typical life expectancy is from 2 to 5 years from diagnosis. However, 20% of people live more than 5 years, 10% more than 10 years. Incidence surveys consistently indicate higher rates of MND amongst males than females, from a ratio of 1.5:1 to 2:1. There is evidence of increasing incidence in a number of countries, including England, USA and all Scandinavian countries. These increases have been attributed to improved methods of diagnosis, increasing awareness of the disease and natural aging of the population (with elder populations being more susceptible) in addition to a true rise in incidence. It appears that MND mortality increases with increasing latitude. ALS is less common in warmer climates, although this could be attributable to lower life expectancies and under-reporting due to reduced awareness of the disease in developing countries. Highest mortalities have been reported in New Zealand, Denmark, Sweden, Switzerland, Finland and Norway.

There is little consensus as to the aetiology of MND. It has been suggested that MND is caused by a slow-acting or latent virus. The very high incidence of ALS observed on the Pacific island of Guam, and on Kii peninsula, Japan, has led to theories about environmental and genetic influences. Other theories include diet, and links with trauma (both physical and psychological) triggering the disease.

3.2 Data

The data were collected by the Neurology Department of the Royal Preston Hospital, in NW England. A total of 128 patients (from a background population of 1.8 million, in Lancashire and south Cumbria, NW England) presented with

MND between 1989 and 1993. A research nursing sister visited each patient (or their relatives) and matched controls at home, and completed a detailed life profile questionnaire, covering demographic, residential, occupational, educational, medical and life-style histories. The data comprise only patients diagnosed within the region. We miss people who are born in, or pass through, the study area, contract MND, but are then resident outside the study area at time of diagnosis. Conversely, we include those residences (within the study region) of patients who have lived largely outside the study area, but have moved relatively recently to Lancashire or south Cumbria and have been diagnosed with MND in Preston.

For reasons outlined above, we have adopted a case-control approach. Each case was matched with two controls: one healthy control and one non-degenerative neurology case (such as Epilepsy). Defining a suitable set of controls is clearly an important, and a non-trivial research problem. We aimed to control for two generally accepted confounding factors, namely age (year of birth within 4 years) and gender. Additionally, a broad matching with area of residence (same District Health Authority) was made, although residence was often more closely matched for the healthy controls, for reasons outlined below. Naturally, only the current residential address could be matched.

The healthy controls were chosen by asking each MND patient or their representative to nominate a friend of the same gender and approximate age. This has introduced room for a degree of over-matching, since friends often share similar socio-economic backgrounds, occupations, diets and hobbies, and, more importantly for the current analysis, would tend to live near each other. This potential tendency for geographical matching can only add weight to our findings however, in that we have adopted a null hypothesis of similar distributions, and any variations from this would be despite a degree of geographical matching. Neurological controls also reported to and were diagnosed by the same neurology department as the MND cases, and hence we should be able to have equal confidence in their diagnosis.

Of the 128 patients, 112 were considered to have a reliable diagnosis and could be successfully matched with two controls. Each of the three matched trios yielded approximately 600 residential moves, 450 of which are within the study region. Preliminary analysis (Mitchell et al, 1995) appears to confirm that the basic characteristics of the data match studies elsewhere, with the male/female ratio being 1.51:1, the mean age at onset being 61 for men and 64 for women, and the average incidence rate per year being 1.42 per 100,000 population.

4. Findings

In Figure 1, below, we demonstrate the distribution of the residences of the MND cases, and note that they 'cluster' in natural population centres such as Blackpool, Preston, Blackburn and Lancaster.

Figure 2 shows the relative risk estimate using the healthy controls with 95% pointwise tolerance contours, and using a bandwidth of 6 Km for the cases and 10.5 Km for the controls. In this figure, darker shades represent higher kernel values; the thick white lines indicate the 97.5% contour of the p-value surface and the thick black lines the 2.5% contour. Recall that points are weighted by length of residence and hence begin to reveal temporal patterns in addition to the more obvious spatial component. This is one aspect which advances our work from that of Bithell (1990) and Gatrell et al. (1991).

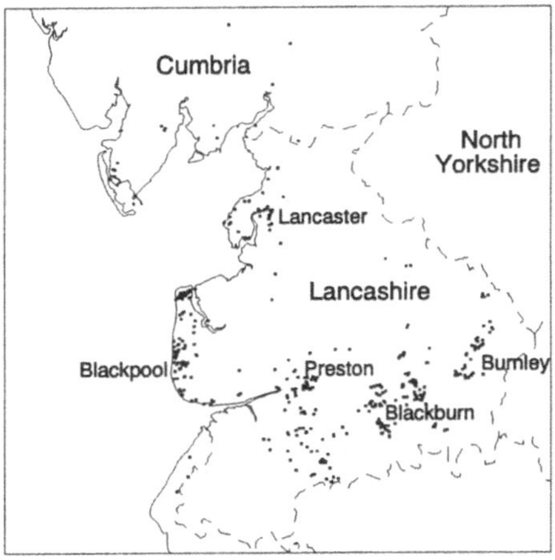

Fig 1. MND patient residences

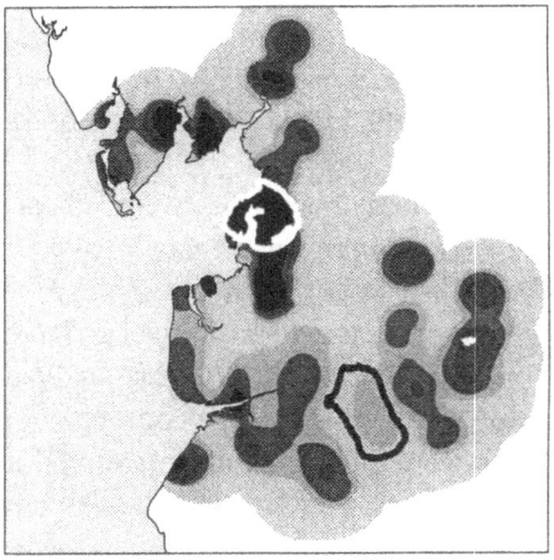

Fig 2. Relative Risk Estimate

It is here that we can begin to discuss apparent patterns. Observe the dark areas in Figure 2, around Lancaster and the larger area between Blackburn and Burnley. Only two statistically significant clusters have been identified - the large Lancaster area and the smaller polygon near Burnley. A large area of significant low relative risk has also been revealed near Blackburn. Whilst these findings do appear interesting, we should be cautious when interpreting them, since we are in an exploratory phase of the analysis. Particular attention should be paid to the various methodological issues highlighted below.

5. Discussion

During our analysis, we noted various problematic anomalies which might be described as 'edge or boundary effects'. These occur when taking the ratio of the two kernels, and in cases where the value of the denominator is small compared to the numerator. This results in strong (but spurious) slivers which translate to peaks which might otherwise represent significant clusters. Only in areas of sparse data does this problem occur. The anomalies we found were also experienced by Bithell (1990). This is obviously an area needing more careful investigation. Are we seeing a form of boundary effect within the study region, where cases and controls intersect? Silverman (1986) touches this issue, but not when specifically referring to the ratio of kernels. The kernel estimator suffers from a drawback when applied to data from long-tailed distributions. Because the window width is fixed across the entire sample, there is a tendency for spurious noise to appear in the tails of the estimates.

If the estimates are smoothed sufficiently enough to deal with the spurious noise, then essential detail in the main part of the distribution is masked. A solution for this is to adopt an adaptive kernel. This is where one adapts the size of the kernel, and hence the amount of smoothing, to the local density of the data. Essentially, this involves altering the bandwidth in response to the underlying point-pattern. This might be a further refinement to our methodology.

Another possible explanation of these so-called anomalies is the nature of the data set. We have less confidence in the data around the periphery of the study area, particularly towards Liverpool and Manchester, to the south of the study region, where one can observe a lack of data. Perhaps this occurs because people living at these locations are not referred to the Royal Preston Hospital, but to another, closer hospital which is outside the study catchment area. We might thus have an incomplete dataset in these areas. Another current concern is that the quantity of both MND patient residences and control residences is simply not enough. If we do not have sufficient density of points, then our kernel estimators do not have enough to work with. An alternative explanation is the nature and spatio-temporal distribution of the controls. We have observed that our methods appear overly sensitive to the distribution of controls in areas of sparse data, where perhaps one or two individual events would alter the kernel surface dramatically.

6. Conclusions

Our findings demonstrate the usefulness of density estimation techniques in providing a useful means of estimating relative risks at different geographical locations. A number of interesting patterns have been observed, which further work will, and needs to elucidate upon, before we can begin to discuss the implications of our findings concerning specific exposure opportunities in the locations identified. The value of our results is emphasised by our desire to remain theory or hypothesis-neutral prior to analysis. Here we consider it worth noting that we have followed and adopted a number of the recommendations proposed by Rushton and Lolonis (1996) in their experimental analysis of birth defect rates, namely:

1. The level of spatial aggregation should be minimised,
2. The search for clusters should begin as 'exploratory spatial analysis' without prior assumptions,
3. The isarithmic maps of relative disease risk should only be used to support further closer investigations, and are not an end in themselves.

A further stage of our work will involve taking time and age slices or windows of the data. This will involve extracting or reselecting parts of the data set, to test ideas about space-time patterning. If the disease distribution was caused by a single environmental event, we would expect to observe temporal and spatial clustering of the patient residences in a specific time slice, at specific locations. Alternatively, if the disease was 'transmitted' by virus, we might observe 'clustering' at specific ages. We could therefore focus our analysis on people aged, say 20-25. This follows on from the work of Riise et al. (1991) in their analysis of residential clustering of Multiple Sclerosis in Norway of patients aged between 13 and 20 years. We must note here that this form of analysis is only now possible due to the recent advancements in GIS software and database query mechanisms.

Acknowledgments

The authors wish to thank Prof. Douglas Mitchell, Consultant Neurologist, Royal Preston Hospital for kindly making the data available.

References

Bailey TC and Gatrell AC (1995) *Interactive Spatial Data Analysis* Longman, Harlow, UK

Bentham G (1988) Migration and Morbidity: Implications for Geographical Studies of Disease *Social Science & Medicine* 26 49-54

Bithell JF (1990) An Application of Density Estimation to Geographical Epidemiology *Statistics in Medicine* 9 691-701

Foster HD (1992) Amyotrophic Lateral Sclerosis: Two for the Price of One? in **Foster HD** (ed) (1992) *Health, Disease & the Environment* Belhaven Press, London

Gatrell AC (1989) On the spatial representation and accuracy of address-based data in the United Kingdom *International Journal of Geographical Information Systems* 3 (4) 335-248

Gatrell AC, Mitchell JD, Gibson HN and Diggle PJ (1991) Tests for spatial clustering in epidemiology: with special reference to motor neurone disease in **Clifford Rose F** (ed) (1991) *New evidence in MND/ALS research* Smith Gordon

Kelsall JE and Diggle PJ (1995) Non-Parametric Estimation of Spatial Variation in Relative Risk *Statistics in Medicine* 14 2335-2342

Löytönen M (1996) *GIS, Time Geography and Health* GISDATA Specialist Meeting, Helsinki, May 1996

Mitchell JD, Davies RB, Al-Hamad A, Gatrell AC and Batterby G (1995) MND risk factors: an epidemiological study in the North West of England *Journal of the Neurological Sciences* 129 (suppl) 61-64

Picheral H (1982) Géographie médicale, géographie des maladies, géographie de la santé *L'éspace géographique* 3 161-175

Riise T, Grønning M, Klauber MR, Barrett-Connere, Nyland H and Albrektsen G (1991) Clustering of residence of multiple sclerosis patients at age 13 to 20 years in Hordaland, Norway *Am Journal of Epidemiology* 133 932-939

Rothman KJ (1990) Keynote presentation. A sobering start for the cluster buster's conference *American Journal of Epidemiology* 132 (suppl 1) S6-13

Rushton G and Lolonis P (1996) Exploratory spatial analysis of birth defect rates in an urban population *Statistics in Medicine* 15 717-726

Schæström A (1996) *Pathogenic Paths? A time geographical approach in medical geography* Lund University Press, Lund, Sweden

Silverman BW (1986) *Density Estimation for Statistics and Data Analysis* Chapman and Hall, London

The use of density estimation techniques in mapping the distribution of Hepatitis A.

Atkinson P.J.[1], Unwin D.J.[2]

1. Public Health Laboratory Service Communicable Disease Surveillance Centre (Thames), 40 Eastbourne Terrace, London W2 3QR, phil@cdscthames.co.uk
2. Department of Geography, Birkbeck College, 7-15 Gressie Street, London W1P 1PA, d.unwin@geography.bbk.ac.uk

Abstract

The application of the method of density estimation to the distribution of hepatitis A in the north west of England is illustrated. Density estimation helps address problems of visualisation and scale in the mapping of disease cases. Several methods are compared and their properties summarised. It is clear that there is no one "correct" method and that users must experiment to find which best suits their data and problem. Current GIS systems struggle in the manipulation and display of the output.

Introduction

There is a long tradition of disease mapping in epidemiology. Some of the earliest examples are distribution maps of yellow fever produced in 1798 (Robinson, 1982). In 1855 John Snow plotted individual cholera cases on map of central London and revealed the key role of contaminated water pumps in spreading the disease (Snow, 1855). Work continued throughout the twentieth century, with a trend in the last few decades towards 'atlases' of disease distribution (Howe, 1970; Gardner *et al.*, 1983; Cliff and Haggett, 1988).

New data and new tools are now available that promise to make disease mapping easier and more powerful. Postcodes can be used to provide detailed locational references for individual cases (Gatrell and Dunn, 1991; Cross, 1990; Bithell, 1990; Openshaw, 1990) and geographical information systems (GIS) can help define the area of interest (Gatrell and Dunn, 1991) and overlay potentially explanatory datasets (Cross, 1990). Spatial clustering methods examine disease concentrated around point or linear features (Openshaw *et al.*, 1987) while space-time contingency tables (Knox, 1964) go a stage further, attempting to determine if disease cases are significantly close in both space and time (Giggs *et al.*, 1980). The underlying risk of disease can be considered using Bayes estimation (Clayton and Kaldor, 1987), or kriging (Oliver, 1994).

Two dimensional density estimation represents a further set of tools now receiving attention (Silverman, 1986; Bithell, 1990; Diggle, 1990; Brunsdon, 1991,1995; Gatrell 1992). It attempts to measure the density of a point distribution at a series of sample points across the area of interest. These sample points are located so that they describe a two dimensional density surface representing spatial variation in the intensity of the spatial process which is assumed to operate. There are a number of problems when attempting to map the distribution and rate of disease, particularly a communicable one. In this paper we examine these problems, and consider how density estimation might be able to help. Various methods of density estimation are considered and compared, to see which is most applicable. The work is based on data from a national collaborative case control study (Maguire *et al.*, 1995) that was undertaken in England to assess the relative contribution of particular risk factors to sporadic hepatitis A in 1990-1.

Problems of disease mapping

Each postcode in Great Britain has been assigned a 100m resolution Ordnance Survey (OS) grid reference (Post Office, 1985; Raper *et al.*, 1992). By placing at dot at the location of the postcode of each case, the disease distribution can be visualised as a simple "dot map" but there are several difficulties with this form of visualisation. First, each postcode in the Postcode Address File which is used to link them to the UK national grid has only one Ordnance Survey grid reference. Whenever multiple cases are found in the same postcode unit, they will be represented only by a single dot. Secondly, each grid reference represents the south-west corner of the grid square within which the first address in the unit postcode lies. This introduces a slight south-westerly bias and means that two postcodes can sometimes be assigned the same grid reference (Gatrell *et al.*, 1991). Therefore, cases in adjoining postcode unit areas with the same grid reference will be represented by only one dot. Communicable diseases have a tendency to cluster, making such multiple representation common. Even when individual cases are assigned individual grid references, the distance between the dots may be so small to make them appear as one. This makes it extremely difficult to visualise the degree of disease intensity in areas with high rates of infection.

Thirdly, the pattern of disease distribution is of little interest without reference to population distribution to establish the at risk population and hence the rate of incidence of disease. At the highest level of spatial resolution, population data in England and Wales are aggregated to census enumeration districts (ED) so that individuals cannot be identified. If no interpolation procedures are used, the ED is the lowest level at which rate can be calculated so the pattern of disease incidence must be analysed using ED boundaries or some higher aggregation of these, leading to difficulties of interpretation arising from the modifiable areal unit (MAU) problem (Openshaw and Taylor, 1987).

Density estimation and epidemiology

The probability density function (pdf) is an important concept in statistics. It provides a natural description of any random quantity X and can be used to determine the probability of X falling above, below, or between particular values. We sometimes have a sample of observed data points from a unknown pdf. Density estimation is the construction of an estimate of the density function from the observed data (Silverman 1986). The best example of a density estimation in one dimension is the familiar histogram, in which we count the number of objects in a series of classes or bins to describe the unknown pdf. The histogram is just one of a number of different density estimation methods that can be applied in one or more dimensions.

Central to all density estimation methods are the concepts of a smoothing parameter and distance decay function. The smoothing parameter defines the area over which each estimate is made. Large smoothing parameter values consider data points far away from the sample point thus producing smooth density estimates, while smaller values consider only the data points close to the sample point so produce a rougher spikier estimate. The distance decay function assigns a "weight" to each data point found in the smoothing parameter defined area. Clearly both these variables will influence the final density estimates, but it has been shown the smoothing parameter has the greatest effect (Gatrell 1992, Bithell 1990). The choice of this smoothing parameter is important, and a large statistical literature has emerged on how this might be accomplished.

Density estimation techniques can be applied to both the pattern of disease incidence and the population data. Two dimensional density methods produce a probability density *surface* showing the distribution of disease. Surface "height" will vary with local disease incidence even if cases are located at the same geographical point by way of the postcode. Visualisation of the density surface thus provides good indicators of the overall variation in true disease density, and detail in areas of high density.

The same two dimensional density estimation methods can be used to distribute population weighted centroid data onto a continuous surface (Bracken and Martin, 1989; Martin and Bracken, 1991). The UK census provides such centroid data, but in principle any zonal aggregated data could have a similar population weighted centroid. Dividing the disease surface by the population surface produces a rate of disease surface which is independent of any fixed set of areal units (Martin, 1989). The modelled values are free from the confidentiality and complexity of individual level data, and avoid the interpretation difficulties given by the MAU problem. It can be argued that density estimate ratios are no better than the "choroplethic" ratios they replace since they vary with the interpolation method in the same way that area-based spatial units are changed. Clearly neither interpolation or area-based approaches are perfect, but we would argue that density estimation methods are likely to give a better measure of the underlying geography of disease incidence (Langford and Unwin 1994).

These advantages have lead various people to apply density estimation techniques to epidemiological problems. Bithell (1990) used them to model the distribution of both childhood leukaemia cases and controls in the vicinity of Sellafield nuclear reprocessing plant in Cumbria and create a risk surface. Diggle (1990) used the kernel estimator density estimation method to produce a density surface of the natural spatial variation in cancers of the larynx in the Chorley-Ribble area. Others have applied these methods to similar point distributions. Brunsdon (1991) examined point distributions of household burglaries and his density surface revealed a 'cliff edge' around a park area where relatively high density housing with a high burglary risk passed into open space with few houses. These examples used the more statistically complex and computationally demanding methods of density estimation, but there are a number of different methods available which vary in their complexity. As a result of this work, it is clear that the influence of the smoothing parameter on the density estimate is well known. Less well known is the influence of the density estimation method which has generally been held to be unimportant. In the next section we examine the distribution of hepatitis A in a small sample area located in North West England (figure 1a). Various density estimation methods are applied and compared in an attempt to assess their impact on the resulting surface form. The basis for all comparisons is purely visual. We acknowledge the subjectivity of this approach, but a fully objective comparison is not possible since the "true" form of the surface is unknown.

A Comparison of Density Estimation Techniques

In order to compare density estimation *methods*, it is important to "fix" the smoothing parameter so that changes in surface form are a result of method and not smoothing parameter. In this study we follow Silverman's suggestion that subjective choice from a range of values is a suitable method (Silverman, 1986). Initially we used a computationally simple estimator (the 'naive estimator') with a range of smoothing parameters producing density surfaces for 233 hepatitis A cases spread over the north-west of England. The results are shown in figure 1(b-d). The 750m parameter produced a spiky surface (figure 1b), where each spike seems to relate to a single postcode containing a case(s). In contrast, the 2km parameter surface was much smoother showing general trends rather than detailed local variation (figure 1d). Next, Brunsdon's cross-validation program (Brunsdon, 1995) was used to calculate the "statistically" optimum smoothing parameter and produced a value of 707m. However, the 750m surface was difficult to interpret visually, complicated by an excessive number of peaks and troughs (figure 1b). As a compromise, a 1km smoothing parameter was adopted for all further work.

To compare the effects of the estimator used, five different density estimation methods were compared; the histogram, naive estimator, kernel estimator, variable kernel estimator, and the nearest neighbour. The histogram, naive and kernel estimators all use fixed smoothing parameters, while the variable kernel and nearest neighbour approaches use variable smoothing parameters. All estimates were calculated using programs written in the C programming language with results written to ASCII files for import into the GIS systems ARC/INFO and GRASS for visualisation. GRASS used the SGD extension designed for dealing with surfaces in 3D, running on a Silicon Graphics machine. Figures in the paper came from GRASS. The output grid resolution was always set at half the value of the smoothing parameter to ensure any variability in the density estimation values would be represented in the surface visualisation. The density estimation equations have not been included here, but are well documented by Silverman (1986).

a) histogram estimator

The histogram estimator is the simplest form of density estimation and, in fact, as "grid square" mapping it has a long history in the cartographic literature (see Langford and Unwin, 1994). The implementation is, of course, almost trivial but examining the shape of the density estimate surfaces, the 1km smoothing parameter histogram surface (figure 2a), with grid squares orientated along the XY axes of the National Grid, shows that it produces a surface remarkably similar to that given by the more complex naive estimator (figure 1c). The level of local detail is not as high, but it forms a good approximation to the overall surface shape. This is a surprising and possibly accidental result as the histogram suffers from the fact that the form of the surface is extremely dependent on the origin and orientation of the smoothing parameter bins.

b) naive estimator

The results of applying the "naive" estimator method of density estimation have already been presented in figure 1. Again, although the computation involved is greater, the principle is easily understood and has been known to cartographers for many years as the "floating grid" technique (Schmid and MacCannell, 1955). In this a fixed size search area defined by the smoothing parameter is located at each sample point in turn over the entire region of interest. At each sample point, with no distance decay function operating, the density estimated is simply a function of the number of data points falling within the search area divided by the size of the search area.

c) kernel estimator

An obvious improvement on the naive estimator is to weight each data point falling into the search area by some function of the inverse of its distance from the sample point. A number of options exist for doing this, but the usual approach centres a known, two-dimensional distance decay function on each sample point in turn across the entire grid and sums for each weighted data point. The kernel estimate should provide a smoother density surface compared to the naive estimator because of the continuity of the distance decay function used. This is not apparent (figure 2b) at this scale. In fact, the kernel estimate seems to have more peaks and troughs around the River Mersey than the naive estimator. This is because nearby data points are actually assigned a higher value with a kernel function than with the naive estimator's weight function, emphasising local detail. If a smaller sample area was considered with less density variation, the advantages of the kernel's smoothing characteristics would be seen more clearly.

d) variable kernel

A major problem with the fixed smoothing in the basic kernel approach is that it may under or oversmooth distributions that themselves vary greatly in density. A smoothing appropriate for a high density region is unlikely also to be appropriate for a low density one. Brundon's (1991) variable kernel approach attempts to address the problem by adapting the smoothing parameter in sympathy with the local point density. It would thus be expected to show both local detail (small smoothing parameter) in areas of high density and regional change (large smoothing parameter) in areas of low density.

The variable kernel approach lives up to this expectation in describing local detail in areas of high density (figure 3a). The Mersey cluster is clearly shown with two principal peaks that are far more apparent than in the histogram, naive or kernel estimated surfaces. However small variations in low density areas are not represented in the surface which remains very smooth.

e) nearest neighbour

A final, adaptive method similar to Brunsdon's is the nearest neighbour method in which the kernel width is allowed to vary by imposition of a minimum number (k) of points to be included criterion. One undesirable property of this technique is that it produces surfaces that integrate to infinity rather than one. This is not a problem when visualizing disease surfaces, but could cause difficulties in calculations of disease rate.

Figure 2 shows the nearest neighbour estimate with k=5 and 7. The Mersey cluster is clearly shown when k=5 as the three data points can be found within a short distance, but the more sparsely distributed cases are not apparent. As k is increased, the surface begins to show regional rather than local trends. When k=7, the sparsely distributed disease cases do not get included, as the nearest neighbour method eliminates minor surface fluctuations in low density areas.

This work suggests that the choice of density estimation method can have a significant impact on the form of the density surface and it is clear that the choice of estimate should depend upon how it is to be used. There are no right or wrong answers. Although fixed smoothing parameter methods work well with the sample HAV distribution, care must be taken with the choice of smoothing parameter. Trying alternatives can actually provide more insight into the data than merely considering a single, automatically produced surface. Good overall impressions are obtained as the same parameters are applied over the whole sample area, although spurious noise can appear in low density areas. Variable smoothing parameter approaches emphasise surface detail in high density areas and eliminate spurious noise in low density areas. These methods are ideal for showing surface breaks and large changes in slope on the density surface, such as around distinct clusters. However, they tend to ignore smaller density fluctuations in areas with few HAV cases. The estimation methods and their characteristics are summarised in table 1.

Once an appropriate density estimation method has been chosen, the results can be used to create a *disease density surface*. In this instance, the naive estimator was selected as it provided a good overall impression of variation in disease density without having large computational overheads. This surface was then divided by a population surface to produce a surface showing rate of hepatitis A (figure 3b). In this area, it shows just two peaks of incidence centred on the urban areas of Liverpool and Tameside where the rates of HAV incidence seem to be anomously high. Unfortunately participation and enthusiasm in the national study varied spatially, making it difficult to interpret these higher than average rates of disease. For example, two additional community nurses were recruited specifically to follow up and record details of all hepatitis A cases in Liverpool throughout 1991, and it might well be that the recorded distribution simply reflects the extent of reporting. With a more reliable data source, density estimation techniques could be used to study the determinants of disease by examining the relationships between relevant socio-economic density estimation surfaces and particular disease density estimation surfaces. The identification of population characteristics which contribute towards disease transmission can help the health-care planner target specific risk groups that may benefit from vaccine.

Conclusion: Problems in producing density estimation surfaces

By their very nature, communicable diseases have extremely discontinuous distributions. This is evident even in the small sample area considered, and it causes a number of problems. Firstly, a clustered point distribution requires a small smoothing parameter to describe the density variation. In order to visualise surface detail the output grid cell size should always be smaller than the smoothing parameter but this results in large output files, many of which contain largely zero density estimates. The difficulties of dealing with such large matrices can be minimised by using any of the usual methods for holding sparse matrices (for example as row, column, value co-ordinates only for "occupied" cells). Secondly, visualisation of continuous, but very rough surfaces, such as those produced for hepatitis A incidence put a great strain on current systems. ARC/INFO had great difficulty displaying the density surfaces, particularly the variable kernel and nearest neighbour. The SGD extension on GRASS was a vast improvement. This is unsurprising considering the Silicon Graphics hardware it uses. Unfortunately such hardware is expensive, so access to such a resource is rare.

In the literature many of the population density surfaces are visualised using grey-scale images. This provides an indication of areas of high and low density, but no information about the degree of density variation within these areas. Other alternatives, such as contouring or "bubble diagrams" (Brunsdon, 1995) suffer from the same problem.

Current GIS output routines seem unable to display and manipulate this type of density surfaces in a satisfactory manner. Neither mesh or continuous surface drapes seem able to expose all the variation in the data. Surface interaction is difficult, discouraging exploratory data analysis and current systems would benefit from many of the tools available in modern scientific visualisation packages.

This work has shown that density estimation techniques should not simply be viewed as 'handle-cranking exercises' as the density estimation method choosen can clearly influence surface form. Choice of smoothing parameter remains important, although the question of the use of the visual optimum or the statistical optimum in producing ratio surfaces remains. Issues concerning visualisation of rough disease density surfaces over large areas have yet to be addressed. Further research and debate is needed.

References

Bithell, J.F. (1990) 'An application of density estimation to geographical epidemiology', Statistics in Medicine, 9, 691-701.

Bracken, I. and Martin, D. (1989) 'The generation of spatial population distributions from census centroid data', Environment and Planning A, 21, 537-543.

Brunsdon, C. (1991) 'Estimating Probability surfaces in GIS: An adaptive technique' in Harts, J., Ottens, H.F.L., and Scholten, H.J. (eds) Proceedings First European Conference on Geographical Information Systems, Amsterdam, EGIS Foundation: Netherlands, 155-64.

Brunsdon, C. (1995) 'Estimating Probability Surfaces for Geographical Point Data: An adaptive kernel algorithm', Computers and Geosciences, 21, 7, 877-894.

Clayton, D. and Kaldor, J. (1987) 'Empirical Bayes estimates of age-standardised relative risks for the use in disease mapping', Biometics, 43, 671-681.

Cliff, A.D. and Haggett, P. (1988) Atlas of Disease Distributions, Blackwell, Oxford.

Cross, A. (1990) 'Using a geographical information system to explore the spatial incidence of childhood cancer in Northern England' in Harts, J., Ottens, H.F.L., and Scholten, H.J. (eds) Proceedings First European Conference on Geographical Information Systems, Amsterdam, EGIS Foundation: Netherlands, 218-29.

Diggle, P. (1990) 'A point process modelling approach to raised incidence of a rare phenomenon in the vicinity of a prespecified point', Journal of the Royal Statistical Society, 153, Part 3, 349-362.

Gardner, M.J., Winter, P.D., Taylor, C.P. and Acheson, E.D. (1983) Atlas of Cancer Mortality in England and Wales 1968-78, Wiley, Chichester.

Gatrell, A.C. (1989) 'On the spatial representation and accuracy of address-based data in the United Kingdom', International Journal of Geographical Information Systems, 3, 335-48.

Gatrell, A.C. (1992) 'Density estimation and the visualisation of point patterns', in Hearnshaw, H. and Unwin, D. (eds) Visualization in Geographical Information Systems. John Wiley & Sons, Chichester.

Gatrell, A.C., Dunn, C.E., and Boyle, P.J. (1991) 'The relative utility of the Central Postcode Directory and Pinpoint Address Code in applications of geographical information systems', Environment and Planning A, 23, 1447-58.

Giggs, J.A., Ebdon, D.S., and Bourke, J.B. (1980) 'The epidemiology of primary acute pancreatitis in the Nottingham Defined Population Area', Transactions of the Institute of British Geographers, 5, 229-42.

Howe, G.M. (1970) National Atlas of Disease Mortality in the United Kingdom, Nelson, London.

Knox, G. (1964) 'The detection of space-time interactions', Applied Statistics, 13, 25-9.

Langford, M. and Unwin, D. (1994) 'Generating and mapping population density surfaces with a Geographical Information System', The Cartographic Journal, 31, 21-26.

Maguire, H.C., Handford, S., Perry, K.R., Nicholas, S., Waight, P., O'Mahony, M., Begg, N.T. (1995) 'A collaborative case-control study of sporadic hepatitis A in England', Communicable Disease Report 5, 3.

Martin, D. (1988) 'An approach to surface generation from centroid-type data', Technical Reports in Geo-Information Systems 5, Computing and Cartography, The Wales and South West Regional Research Laboratory.

Martin, D. and Bracken, I. (1991) 'Techniques for modelling population-related raster databases', Environment and Planning A, 23, 1069-1075.

Oliver, M.A. (1994) 'Geostatistics, rare diseases and the environment' in Scholten, H., Fischer, M. and Unwin, D. (eds)(in press) Spatial Analytical Perspectives on GIS in the environment and socio-economic sciences. Taylor and Francis, London.

Openshaw, S. (1990) 'Automating the search for cancer clusters: a review of problems, progress and opportunities', in Thomas, R.W. (ed) Spatial Epidemiology, 48-78, Pion, London.

Openshaw, S., Charlton, M., and Cross, A.E. (1990) 'Building a prototype Geographical Correlates Exploration Machine', International Journal of Geographical Information Systems, 4, 297-311.

Openshaw, S., Charlton, M., Wymer, C. and Craft, A.W. (1987) 'A Mark I geographical analysis machine for the automated analysis of point data sets', International Journal of Geographical Information Systems, 1, 335-58

Openshaw, S., and Taylor, P.J. (1987) 'The Modifiable Areal Unit Problem' in Wrigley, N. and Bennett, R.J. (eds) Quantitative Geography: A British View, Routledge, London.

Post Office (1985) 'The postcode address file digest', The Post Office, London.

Raper, J., Rhind, D.W., and Shepherd, J. (1992) 'Postcodes: The New Geography', Longman, London.

Robinson, A.H. (1982) Early Thematic Mapping in the history of Cartography. University of Chicago Press, Chicago, pages 170-82.

Schmid, C.F. and MacCannell, E.H. (1955) 'Basic problems, techniques and theory of isopleth mapping', Journal of the American Statistical Association, 50, 220-239.

Snow, J. (1854) On the mode of Communication of Cholera, 2nd edn. Churchill Livingstone, London.

Silverman, B.W. (1986) Density Estimation, Chapman and Hall: London.

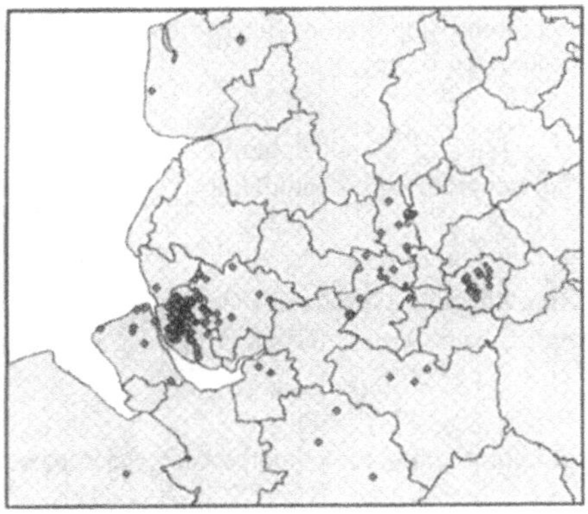

Figure 1a
Location of hepatitis A cases

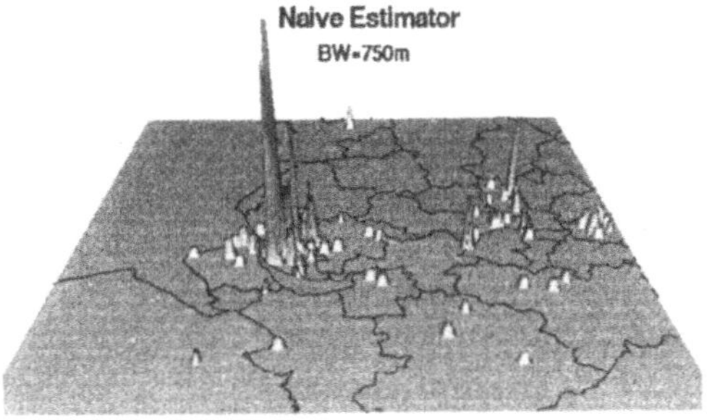

Figure 1b
750m naïve estimator surface

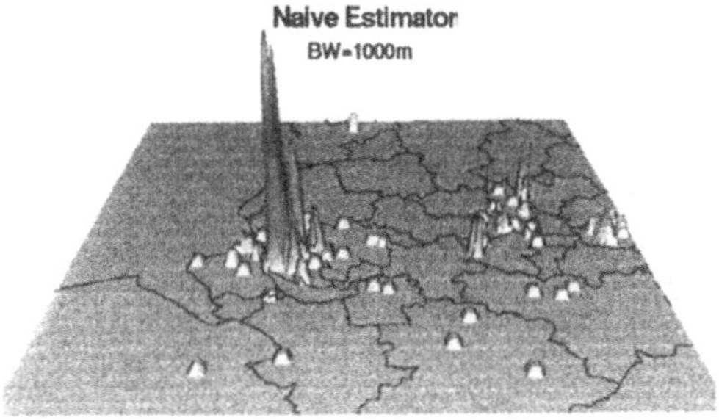

Figure 1c
1000m naïve estimator surface

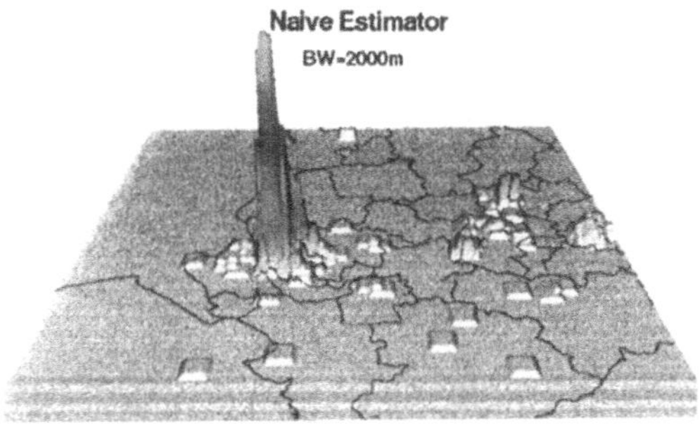

Figure 1d
2000m naïve estimator surface

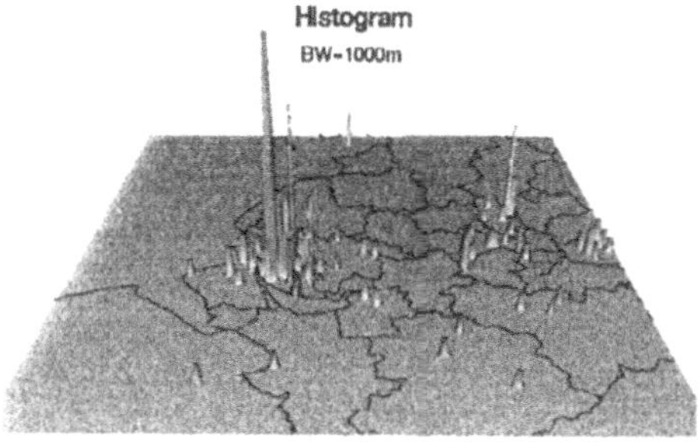

Figure 2a
1000m histogram surface

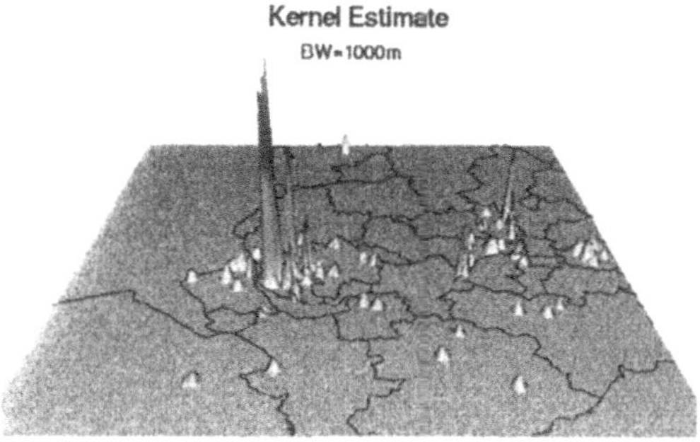

Figure 2b
1000m kernel estimator surface

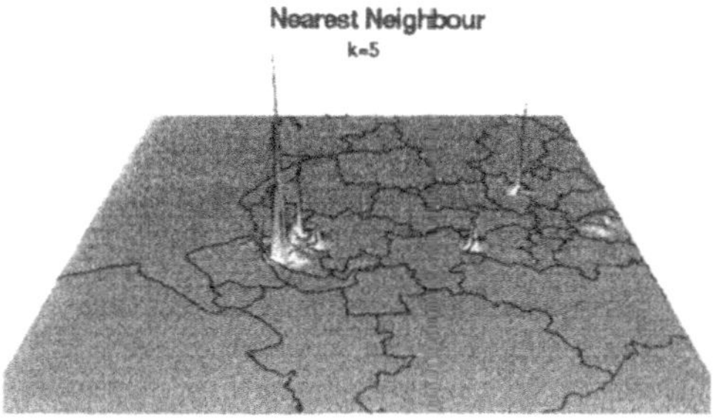

Figure 2c
k=5 nearest neighbour
estimator surface

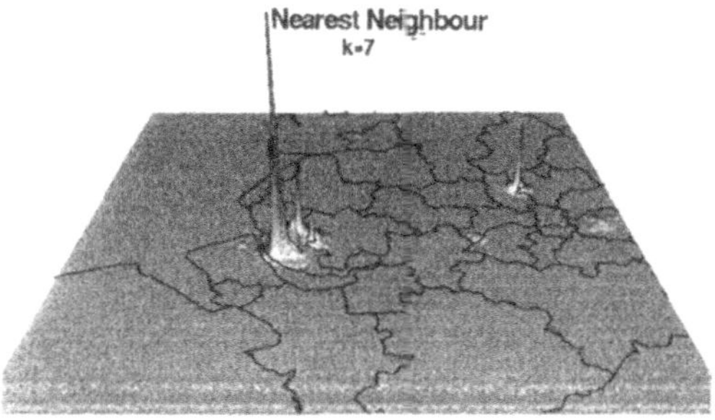

Figure 2d
k=7 nearest neighbour
estimator surface

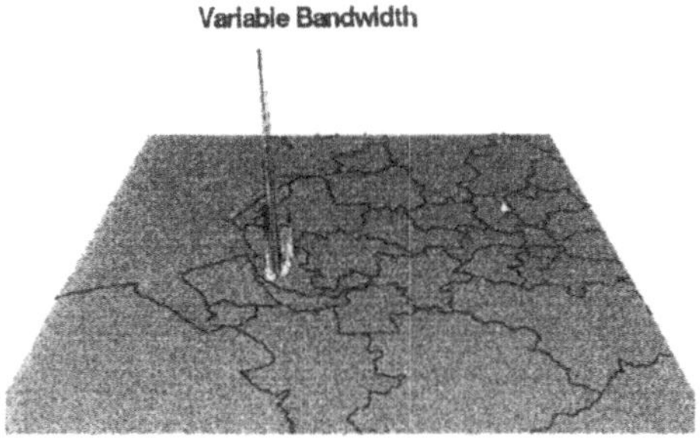

Figure 3a
Variable kernel estimator surface

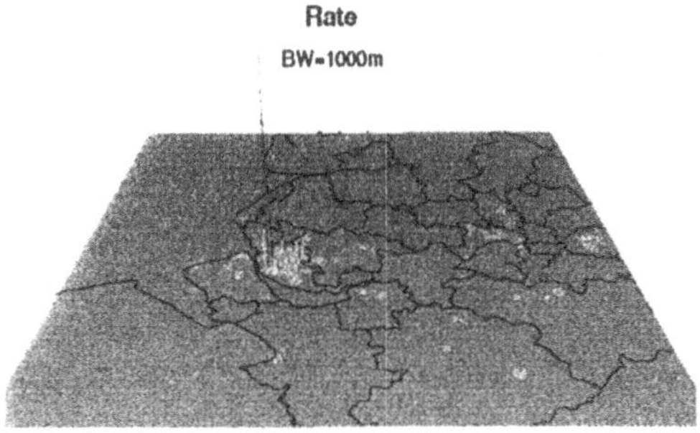

Figure 3b
1000m hepatitis A
estimated rate surface

FIXED smoothing parameter (histogram, naive estimator, kernel)	**VARIABLE smoothing parameter** (variable kernel, nearest neighbour)
Choice of smoothing parameter important	Choice of initial smoothing parameter important (particularly if used as final smoothing parameter)
Creates detailed density surface in areas of high and low density	Creates very detailed density surface in areas of high density. Eliminates "noise" in areas of low density
Good overall impression of underlying distribution	Emphasises breaks in the density surface
	Computationally expensive
THIS evidence shows type of FIXED smoothing parameter method unimportant	THIS evidence shows type of VARIABLE smoothing parameter method unimportant, depending on integration needs

Table 1. Summary of fixed and variable density estimation characteristics.

Aggregation Levels in Visualisation and Spatial Analysis of Epidemiological Data

Nanja van den Berg, Ramona Rudolph and Karl-Rainer von der Ahé

ZRI e. V. - Spatial Data Centre
Brandteichstraße 19
17489 Greifswald

Fon: +49 3834 550 504
Fax: +49 3834 550 430
e-mail: zri@uni-greifswald.de

Abstract

This paper describes the visualisation and the analysis of spatial medical data on different aggregation levels. We intend to give an overview of the behaviour of some methods on different generalisation levels, illustrated by some examples.

1. The Use of GIS in Epidemiology: An Introduction

The use of Geographical Informationsystems (GIS) in epidemiology is an effect of the interdisciplinary co-operation between epidemiologists and geographers in the research field called medical geography, geomedicine or geographical epidemiology. These terms can be defined (among others) as the description and explanation of spatial patterns of disease incidence or prevalence and mortality.

The development of GIS since the seventies has, together with the increased achievements of computers, improved the possibilities to visualize and analyse spatial medical data, especially if large datasets are involved. One of the benefits of GIS is the ability to construct and execute more or less complex database queries which can be visualized in a map. It is for example possible to extract cases with special characteristics, like site, sex or age [1]. A more complex query is the combination of such characteristics with for instance environmental data (overlay operations).

This paper focuses on difficulties in visualising and analysing medical data, due to the choice of suitable aggregation (or generalisation) levels and the georeferencing following from this. The background of this paper is a research project called ERIS (Development of a Epidemiological Spatial Information System), which is an interdisciplinary project to develop a GIS-application, suiting to the needs and ideas of epidemiologists.

2. Data Collection and Georeferencing in Epidemiology

Actually, the title of this section is inconsistent as data collection in a spatial epidemiological context is rare. In most cases, medical data are collected for other but not for spatial purposes. A lot of difficulties in georeferencing medical data arise from data collection without a thought of later spatial research.
The collection and availability of medical data is different across Europe. Some countries use censuses as data source (for example the United Kingdom, Italy and Spain), others use registers (Denmark, Finland) or surveys (Germany). A few countries use a combination of methods, like the combination of registers and surveys (the Netherlands) or registers and census (Norway and Sweden) [2].
Often, data are collected on community or postcode level. Any research below this level is difficult. In some cases, data on a more detailed aggregation level are available through surveys like ISAAC (International Study of Asthma and Allergies in Childhood) [3]. Summarised, it can be said that the limitation in using GIS, caused by data protection acts and data collection for initially non-spatial purposes, is a limitation in the choice of aggregation level.
Another limitation, caused by the used data is spatial inconsistency in the datasets. Especially in datasets surveyed without co-operation with GIS-specialists, spatial inconsistency can 'pollute' the data in such a way, that renders georeferencing difficult [4].

3. The Level Makes a Difference

As shown in section 2, the choice of the generalisation level is in many cases a unconscious or forced choice. Unfortunately, the level of generalisation has an important influence on spatial patterns, as the next two sections show both in visualisation and in analysis.

3.1 Aggregation Levels and Visualisation

Depending on the kind of the spatial analysis procedure, visualisation of data is the first and/or last step in both epidemiological and other spatial investigations. To start a spatial analysis, a visualisation of the raw or only slightly adjusted data can be very useful to obtain a first overview of the spread of data and possible spatial patterns [5]. Examples for such investigations are atlases of diseases, for example the cancer atlas of the federal state of Nordrhein-Westfalen in Germany

[6]. Usually, atlases of diseases focus on the visualisation of medical data, statistical analyses to prove the significance of spatial patterns are rare.
To visualize the results, the output of spatial investigations can be a thematic map. Here, spatial investigations are not only statistical analyses, but also complicated database queries. Ideal is the possibility of instant mapping, which is the construction of the thematic map after an analysis or database querie in only a few steps.
An interesting question is, which geographical unit is most appropriate in representing epidemiological data. The choice of the size of the visualisation unit is crucial: the number of cases of diseases is usually small, therefore, too small visualisation units are unsuitable. On the other hand, units must be small enough to recognise local environmental or social effects on the disease rates [7,8].
The choice of the visualisation unit is dependent on data protection acts and nature, density, and geographical reference of the data. If spatial information is available on the level of xy-coordinates, point maps can be made (fig. 1). Although point maps are the most detailed way to visualize data, it is only an appropriate unit if both cases and controls are available. Controls provide some knowledge about population density. Without controls, it is impossible to distinguish between the spatial patterns of population density and cases of diseases.
Usually, epidemiological data will be visualized in small areas, for example grids, postcode areas or any administrative units (figs. 2 and 3). Small-area data can be visualized in thematic maps using different statistical measures. Often, incidence or prevalence rates are calculated. The disadvantage of this method is that possible local differences between the areas are covered by variation in the data: extreme incidence rates of diseases appear in areas with a low population density and rare diseases show more variation than more common diseases.
Another possibility to visualize small-area data is the calculation of significance levels. Now, the opposite effect appears: extreme significance levels tend to be in areas with large populations and common diseases show more spatial variability than rare diseases [9].
Smoothing methods can reduce the spatial variability by spatial averaging the data. Figure 4 shows the result of a smoothing method, where the disease rate of a gridcell is dependent on the disease rates, distance and population density of his neighbour cells, the edge effects are not taken into account with this method. The smoothed map shows a less 'speckled' pattern as figure 3.
The disease rate of a gridcell results from the weighted sum of the disease rates of his neighbourcells till a certain distance:

$$\mathrm{MR}=\sum_i w_i MR_i \tag{1}$$

w_i is the weight, MR_i is the age-standardized disease rate from gridcell i. The weights are calculated with (2):

$$w_i \propto \exp\left(-\frac{1}{2\sigma^2} d_i^2\right) \cdot b_i \qquad (2)$$

where σ is a parameter to control the influence of distance d. b_i is the population density of gridcell i. This simple method to smooth data is for example used in the cancer atlas of the federal state of Nordrhein-Westfalen in Germany [6].
A more complicated method of smoothing spatial data is the use of empirical Bayes estimates. The Bayesian method is a weighted compromise between standardised rates, the overall mean relative rate and a local mean of the relative rate in nearby areas [9].

3.2 Aggregation Levels and Spatial Analysis

3.2.1 An Overview

Analog to the subject of aggregation in cognitive mapping [10], there are three ways to implement aggregation into the analysis process:

1. the method of *disaggregation* is actually no aggregation. The data is analyzed and visualized individually on point level;
2. *collective aggregation* is a method in which the data is aggregated prior to analysis; the result of the analysis represents the whole dataset;
3. the third way to aggregate data is the so-called *individual aggregation* method. Here, the data are analyzed individually on point level, but the results are visualized pooled and averaged [10].

Spatial analysis in epidemiology means often the search for clusters. Clusters are any localised aggregations of cases in space and /or time. As in visualisation, the results of clusteranalysis depend heavily on the used system of areal units. Techniques of clustering may fail to detect 'real' clusters which pass one or more administrative boundaries [11] in a small-area-system or which are too small to be detected in a large-area-system.

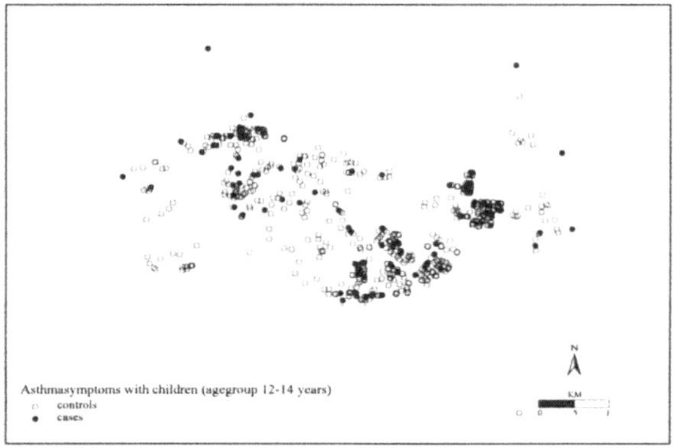

Fig. 1. Asthmasymptoms, point map

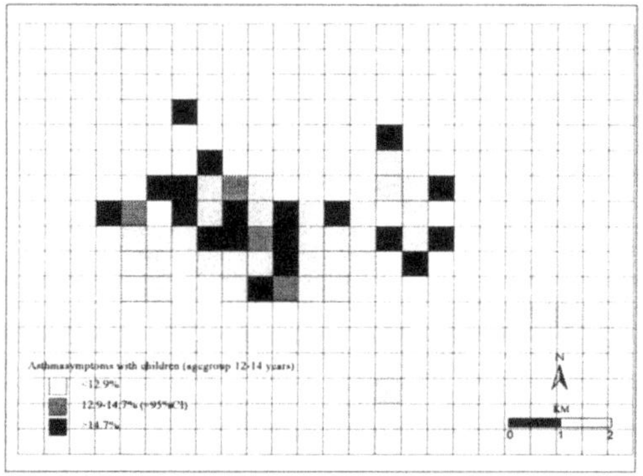

Fig. 3. Asthmasymptoms, 500m grid

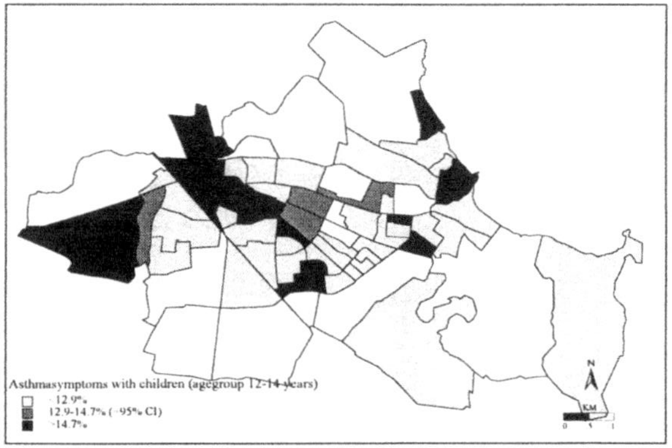

Fig. 2. Asthmasymptoms, communal statistical districts

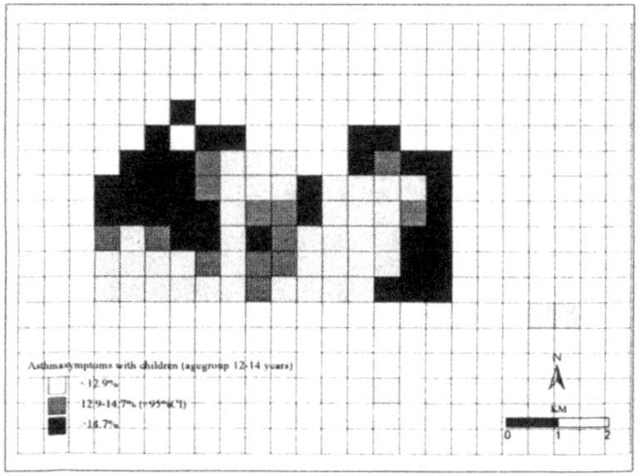

Fig. 4. Asthmasymptoms, 500m grid, smoothed

In principle, there are two ways to carry out a spatial epidemiological investigation: the first way is a general investigation: an investigation without previous knowledge about any possible causes for any spatial patterns. Statistically, this is the most objective and reliable method.
The second possibility is a focused investigation, which is an investigation with previous knowledge or suspicion, often based on public concern. A well-known example for the second method was the Black investigation following public concern about a high number of cases of childhood leukaemia around the Sellafield nuclear plant in the north-west of England [12]. Focused investigations concentrate in most of the cases on the statistical evidence of suspected spatial patterns.
The difference between focused and general tests lies in their null hypotheses: the null hypothesis of a general test says that the distribution of cases is random through the research area. The null hypothesis of a focused test differs slightly as it says that there is no spatial clustering around the focused point source (see also [13]). With other words: focused tests have a defined starting point, for example the suspicion that exhaust emissions are associated with an increased childhood asthma risk.
Both investigation methods have in common, that it is very difficult to construct an objective and significant picture of the actual health situation in any research area. Due to this problem are severe data protection acts that render detailed georeferencing difficult, low disease incidences which make statistics less reliable and extreme differences in the underlying population density, which can cause false clusters.
For different techniques of clustering I refer among others to Diggle et al [14], Openshaw [15], Jaquez [16] and the CDC Guidelines [17].

3.2.2 Analysis of Epidemiological Data on Different Aggregation Levels

To test the effect of aggregation, we analyzed data (restricted to the city of Greifswald) of the West-Pomerania part of the international ISAAC. ISAAC is a worldwide study, which was carried out in 48 countries all over the world [3] The West-pomerania part of the ISAAC-data was collected by the Institute of Hygiene and Environmental Medicine of the University of Greifswald.
The datasets, available for ERIS, exist in form of the records for about 6000 children: 3000 children in the age of 12-14 years filled in a questionnaire (1025 in the city of Greifswald), 3000 parents did the same for their 6-8 years old children (950 in the city of Greifswald). The children and parents were mainly asked for skin and breath complaints, state and site of the houses, the number of cars driving past their houses and the presence of waste gases. The symptoms under investigation

were described in the questionnaires and some symptoms were, concerning the agegroup 12-14 years , shown on video.
From this datasets, we derived nine new datasets for each agegroup, restricted for the city of Greifswald: for three diseases (moderate-heavy asthma, moderate-heavy hayfever and moderate-heavy eczema) and three aggregation levels (point level, 250m-grid and 500m-grid).
The datasets were analyzed with three methods for general clusteranalysis: Cuzick and Edward's method on point level (with regard to 3.2.1: disaggregation method), Grimson's method and Moran's I adjusted for population size (I_{pop}) on small area levels (with regard to 3.2.1: collective aggregation method). As a reference, we used test power investigations of Oden et al [18]. The power of a statistical test is measured as the chance to detect real clusters in a dataset. Oden et al concluded, under the restriction that the methods aren't strictly comparable because of different data requirements, that the I_{pop} test has very good power, Cuzick-Edwards test has good til moderate power, dependent on the simulation parameters, and Grimson's test is the least powerful method of those three. However, the choice of an analysis method is dependent on the geographical and epidemiological characteristics of the dataset.
Cuzick and Edward's method is suitable for datasets where exact spatial locations of cases and controls are known. The test statistic T_k is the count of the number of cases whose nearest neighbours are cases and not controls [19]:

$$T_k = \sum_{i=1}^{N} \delta_i d_i^k \qquad (3)$$

$\delta_i = 1$ if observation i is a case and 0 if it is a control;
$d_i^k = 1$ if the kth nearest neighbour to i is a case, 0 otherwise.

A detailed description of Cuzick and Edward's test is given by Cuzick and Edward's [20].
Grimson's test has a broad range of applications: it can be used for general cluster analysis in space, time and space-time. Here, we use the test to detect clusters in small areas in space. The test statistic is the count of the number of pairs of labeled objects adjacent to another [19]. In this case, labeled objects are areas with rates of disease, higher than the 95% border of the confidence interval. Grimson's test doesn't take the population density into account. See for a description [19].
The third used method is I_{pop} (Moran's I adjusted for population size). It is a test for spatial autocorrelation in disease rates that accounts for geographic variation in

population size [18]. We performed this test for small areas on the same generalisation levels as for the analysis with Grimson's method.
The used generalisation levels for the I_{pop} and Grimson's test were 250- and 500m.-grids. The choice of the grids was based on the rule of thumb that we needed 5 cases in every gridcell. Unfortunately, the population density in the city of Greifswald is very irregular, so it was impossible to find an acceptable grid with a minimum of 5 cases per gridcell. The 250- and 500m.-grids came near this value.
We performed both I_{pop} and Grimson's test with four and with eight neighbours (Fig.5). The results of the clusteranalyses are summarized in tables 1-3.

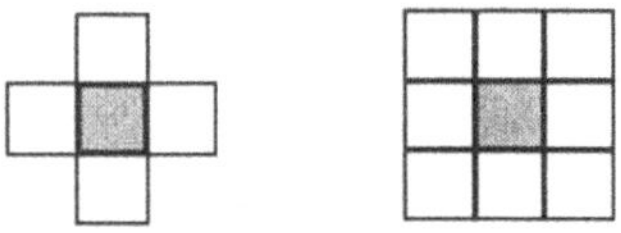

Fig. 5. Gridcell with four (a) and eight (b) neighbours

Table 1 shows the results of global clusteranalysis on point level with the Cuzick and Edwards test. Here, no global clustering was detected. Table 2 shows the results of global cluster analysis with Grimson's test on a 250m- and a 500m-grid. Using Grimson's test, we find no clustering for asthma, and clustering in both agegroups for hayfever and eczema.
Table 3 shows the results of I_{pop} at the same aggregation levels. Here, we have clustering for hayfever (agegroup 12-14 years, 250m-grid) and asthma (agegroup 6-8 years, 250m-grid).
Grimson's test produces more significant clusters as I_{pop}, probably because of clusters, due to the underlying differences in population density. Interesting ist the fact, that I_{pop} finds significant clustering for asthma in the youngest agegroup (eight neighbours), where the normally less powerful Grimson test finds no clustering [18].
From the aggregation point of view, there are clear differences in the results between the three aggregation levels. The point level shows no clustering at all. Using Grimson's test, we find both in the 250m- and in the 500m-grid six cases of clustering. Except for two cases (hayfever in the youngest and eczema in the second agegroup), the results are the same in both grids. Further, the cases of clustering in the 250m grid are the same in both the four- and eight-neighbour variation.
These results differ from the results using I_{pop}. Here, we find no clustering in the 500m-grid. For hayfever in the agegroup 12-14 years, we find clustering in the

250m-grid in both the four- and eight-neighbour variation of the test. Asthma shows clustering in the 250m-grid in the agegroup 6-8 years.

Table 1: p's for Cuzick-Edwards (I=agegroup 6-8 years, II=agegroup 12-14 years). Bold: significant p when $\alpha=0.05$

	Points, Bonferonni	Points, Simes
Asthma I	1.0000	0.2616
Asthma II	1.0000	0.8131
Hayfever I	1.0000	0.4077
Hayfever II	1.0000	0.4604
Eczema I	1.0000	0.4460
Eczema II	1.0000	0.7760

Table 2: p's for Grimson (I=agegroup 6-8 years, II=agegroup 12-14 years) Bold: significant p when $\alpha=0.05$

	250m-grid, 4 neighbours	250m-grid, 8 neighbours	500m-grid, 4 neighbours	500m-grid, 8 neighbours
Asthma I	0.591	0.419	0.680	0.489
Asthma II	0.284	0.325	0.898	0.883
Hayfever I	**0.009**	**0.0009**	**0.042**	0.093
Hayfever II	**0.00034**	**0.00098**	**0.015**	**0.036**
Eczema I	**0.032**	**0.003**	**0.048**	**0.044**
Eczema II	0.181	0.134	0.149	**0.017**

Table 3: p's for I_{pop} (I=agegroup 6-8 years, II=agegroup 12-14 years) Bold: significant p when $\alpha=0.05$

	250m- grid, 4 neighbours	250m- grid, 8 neighbours	500m- grid, 4 neighbours	500m- grid, 8 neighbours
Asthma I	0.072	**0.015**	0.147	0.051
Asthma II	0.955	0.687	0.798	0.722
Hayfever I	0.266	0.251	0.677	0.316
Hayfever II	**0.010**	**0.010**	0.139	0.068
Eczema I	0.906	0.694	0.304	0.659
Eczema II	0.203	0.113	0.216	0.516

The results of this analysis are the basis for further, focused, cluster analysis. In case of general clustering, we will try, in an intensive co-operation with epidemiologists, to put forward hypotheses about possible point- or linesources as causes for clusters. After developing hypotheses, we will try to analyse the data with methods for focused clustering (for example Stone [21] or Besag and Newell [22]) with these point- or linesources as a starting point.

4. Concluding Remarks

The use of GIS in medical research is an interesting, interdisciplinary field of research. Visualisation is a great help to represent spatial medical data, spatial analysis methods can be used to find significant differences from the expected spatial patterns of diseases.
Due to limitations in the availability of spatial medical data, visualisation and spatial analysis are often restricted to certain aggregation levels. The choice of aggregation level has a large impact on the results of both visualisation and spatial analysis. The result is, that spatial patterns of diseases may be undiscovered, because of unsuitable spatial research units. More research is needed, to find the effects of different aggregation levels on various methods of spatial analysis.

Acknowledgements

This work is funded by a grant from the Ministry of Education and Cultural Affairs of the Federal State of Mecklenburg-Vorpommern, Germany. We would like to thank Geoffrey Jacquez (Biomedware, Inc.) for putting the analysis software STAT! at our disposal and for discussing the methods of statistical analysis, Antje Sdanawitschus for preparing data for analysis, Anett Spring and Andrea Voelkner for carefully reading the manuscript.

References

[1] **Gatrell, A. C. and C. E. Dunn** (1991) GIS in Epidemiological Research: Analysing Cancer of the Larynx in North West England. North West Regional Research Laboratory Lancaster University, Research Report No. 12.

[2] **Lovett, A., R. Haynes, G. Bentham, S. Gale and J. Brainard** (1996) Improving Health Needs Assessment through Utilising Patient Register Information in a GIS. Paper presented at the Helsinki GISDATA meeting, May 1996. To be published in: Gatrell, A. C. and M. Löytönen (eds.). GIS and Health. Taylor and Francis [1997].

[3] **Asher, M. I., U. Keil, H. R. Anderson, R. Beasley, J. Crane, F. Martinez, E. A. Mitchell, N. Pearce, B. Sibbald, A. W. Stewart, D. Strachan, S. K. Weiland and H. C. Williams** (1995) International Study of Asthma and Allergies in Childhood (ISAAC): Rationale and Methods. European Respiratory Journal, 8, 483-491.

[4] **Berg, N. J. van den und K.-R. von der Ahé** (1997) Geoinformationssysteme in der Epidemiologie. Kartographische Nachrichten 2: 52-59.

[5] **Ahé, K.-R. von der and N. J. van den Berg** (1997) The epidemiological GIS-Application ERIS. In: Proceedings of the JEC-GI, Vienna, 16-18 April 1997; 833-842.

[6] **Pesch, B., U. Halekoh, U. Ranft, M. Richter und F. Pott** (1994) Atlas zur Krebssterblichkeit in Nordrhein-Westfalen. Ministerium für Arbeit, Gesundheit und Soziales des Landes Nordrhein-Westfalen, Düsseldorf.

[7] **Möhner, M. und R. Stabenow** (1994) Kartographische Darstellung epidemiologischer Sachverhalte und Methoden zur Clustererkennung. In: Heinemann, L. (ed.) Epidemiologische Arbeitsmethoden. Fischer, Jena.

[8] **Bithell, J. F.** (1992) Statistical Methods for Analyzing Point-source Exposures. In: P. Elliott, J. Cuzick, D. English and R. Stern (eds.). Geographical and Environmental Epidemiology. Oxford University Press.

[9] **Clayton, D. and L. Bernardelli** (1996) Bayesian Methods for Mapping Disease Risk. In: P. Elliott, J. Cuzick, D. English and R. Stern (eds.). Geographical and Environmental Epidemiology. Oxford University Press.

[10] **Kitchin, R. and A. S. Fotheringham** (1997) Aggregation Issues in Cognitive Mapping. Professional Geographer, 49 (3): 269-280.

[11] **Merril, D. W., S. Selvin and M. S. Mohr** (1992) Density Equalizing Map Projections: Techniques and Applications. Paper presented at the Workshop on Statistics and Computing in Disease Clustering, Stony Brook NY, 23-24 July 1992.

[12] **Gardner, M. J.** (1992) Childhood Leukaemia Around the Sellafield Nuclear Plant. In: Elliot, P., J. Cuzick, D. English and R. Stern (eds.): Geographical and Environmental Epidemiology. Oxford University Press.

[13] **Jacquez, G. M., R. Grimson, L. A. Waller and D. Wartenberg** (1996) The Analysis of Disease Clusters, Part II: Introduction to Techniques. Infection Control and Hospital Epidemiology 17: 385-397.

[14] **Diggle, P. J., A. C. Gatrell and A. A. Lovett** (1990) Modelling the Prevalence of Cancer of the Larynx in Part of Lancashire: A New Methodology for Spatial Epidemiology. In: R. W. Thomas (ed.). Spatial Epidemiology. Pion, London.

[15] **Openshaw, S., M. Charlton, C. Wymer and A. W. Craft** (1987) A Mark 1 Geographical Analysis Machine for the Automated Analysis of Point Data Sets. International Journal of Geographical Information Systems 1: 335-338.

[16] **Jaquez, G. M.** (1996) A k-Nearest Neighbor test for Space-Time Interaction. Statistics in Medicine, Vol. 15:

[17] Guidelines for Investigating Clusters of Health Events (1990). MMWR 39 (RR-11), 1-23. CDC, Atlanta.

[18] **Oden, N., G. Jacquez and R. Grimson** (1996) Realistic Power Simulations Compare Point- and Area-Based Disease Cluster Tests. Statistics in Medicine, Vol. 15: 783-806.

[19] **Jacquez, G. M.** (1994) Stat! - Statistical Software for the Clustering of Health Events. Biomedware, Ann Arbor.

[20] **Cuzick, J. and R. Edwards** (1990) Spatial Clustering for inhomogeneous Populations. Journal of the Royal Statistical Society, B 52: 73-104.

[21] **Stone, R. A.** (1988) Investigations of Excess Environmental Risks around Putative Sources: Statistical Problems and a Proposed Test. Statistics in Medicine, Vol. 7: 649-660.

[22] **Besag, J. and J. Newell** (1995) The Detection of Clusters in Rare Diseases. Journal of the Royal Statistical Society, Series A, 154: 143-155.

Statistical Methods

The Detection of Outbreaks of Infectious Disease

C. P. Farrington and A.D. Beale

PHLS Statistics Unit
Communicable Disease Surveillance Centre
61 Colindale Avenue
London NW9 5EQ

e-mail addresses:
pfarring@phls.co.uk   *abeale@phls.co.uk*

Abstract

Over the past decades, computerised national and multi-national surveillance systems for infectious diseases have grown in importance and versatility. A common feature of such systems is the large quantity of data they generate, hence the need for a degree of automation involving more formal statistical approaches. However, while considerable methodological advances have been made in the detection of temporal clusters of disease, rather less attention has been devoted to the statistical issues involved in the prospective detection of outbreaks.

In this paper we discuss some of the methodological problems of prospective outbreak detection, and review some of the statistical methods which have been developed or may be adapted for this purpose. We conclude that automated detection systems can provide a 'safety net' which supplements, but cannot replace, other methods of surveillance, and emphasize the need for close inter-disciplinary collaboration.

1. Introduction

In the closing years of this century, the great advances in the prevention and control of infectious diseases through improved sanitation and mass vaccination may be contrasted with the new public health challenges posed by the industrialisation and globalisation of food production. Thus, in the United Kingdom, considerable public concern has resulted in recent years from the increased incidence of *Salmonella enteritidis* PT4 infections associated with egg consumption, *Escherichia coli* O157 infections from contaminated meat, and the appearance of a new variant of Creutzfeldt Jakob Disease believed to be linked to cattle infected with Bovine Spongiform Encephalopathy. In this context, the prevention and control of outbreaks of infectious disease remains one of the priorities of public health policy.

This paper addresses the methodological and statistical aspects of outbreak detection. In contrast with many other areas of epidemiology in which statistical methodology has come to play a central role (for instance, clinical trials, survival analysis, study design, case-control studies), formal statistical methods play only a minor part in the detection of infectious disease outbreaks. Conversely, in contrast to an extensive statistical literature on the *retrospective* detection of temporal and spatial clusters of disease, rather less attention has been devoted to the statistical problem of *prospective* outbreak detection.

Nevertheless, the computerisation of surveillance systems and the large amounts of data on thousands of different infectious agents and pathogens which such databases generate, together with the ready availability of substantial computing power, may provide the impetus for further work in the area of computer-assisted outbreak detection.

We begin by discussing some of the methodological problems which a detection system for infectious diseases must address, including the problems of reporting delays, systematic variation, past outbreaks, robustness and false detections.

We then review the statistical methods which have been proposed for, or which may be adapted to, outbreak detection. These include regression methods, time series, and methods inspired by statistical process control techniques. We also discuss the contribution of spatial statistics to some aspects of outbreak detection.

Finally, we briefly comment on the contribution to outbreak detection of computer-based, automated systems.

2. Methodological Problems

An outbreak is often defined as two or more cases of infection by a common agent which are epidemiologically linked in some way, for instance by exposure to the same contaminated foodstuff. This definition is of little practical relevance for identifying outbreaks, as it presupposes that detailed epidemiological information is available on the cases. The aim of outbreak detection is to identify unusual temporal clusters of cases which are then investigated in greater detail. Thus, in practice, an outbreak is defined operationally as an increase in the number of cases beyond expected levels.

Figure 1: Reports of Salmonella typhimurium DT104 England and Wales, 1990-1997

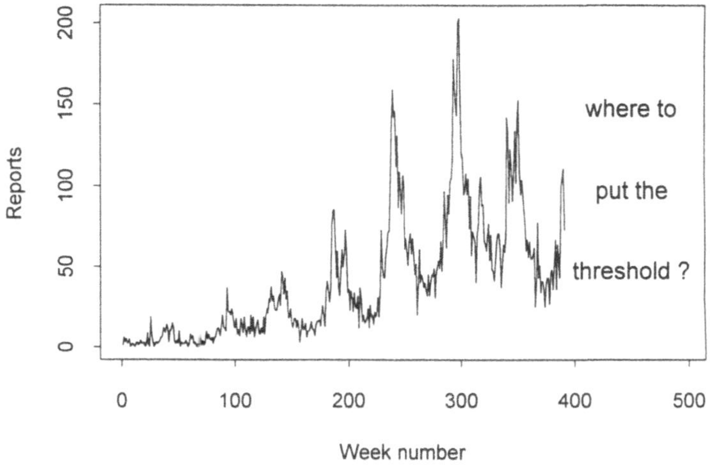

This definition requires the specification of a *baseline* level of reports, and a *decision mechanism* to judge whether current counts significantly exceed this baseline. Accordingly, most outbreak detection systems involve the calculation of threshold values which depend on baselines, these being derived from historical data. If an organism count exceeds the threshold, this may trigger further epidemiological investigations to determine common epidemiological factors and, if an outbreak is confirmed, to introduce control measures.

Stated in these terms, the problem of outbreak detection is a simple one, graphically summarised in figure 1. However developing an effective system is a challenging task. In this section we discuss some of the methodological problems which outbreak detection systems must engage with.

2.1 Reporting Delays

One of the major differences between retrospective and prospective outbreak detection is the issue of reporting delays. Perhaps surprisingly in view of its fundamental importance, this issue is rarely discussed. Retrospective outbreak detection is usually undertaken on data ordered by date of event, ideally date of infection, or date of onset or specimen collection if infection date is not available.

As is standard practice in most statistical analyses, the data are validated, errors are corrected, missing fields entered and so on.

In contrast, for prospective detection, the data must be analysed as it accumulates: there is minimal time, if any, for any data validation. Furthermore, the data are subject to reporting delays whereby the most recent infections are less likely to be reported. This results in an underestimate of current and recent counts when ordered by date of infection or specimen. Clearly, this represents a major obstacle, since the most recent data is precisely that upon which outbreak detection is based.

There appear to be two approaches to handling this problem. The first is to apply a correction factor to recent data, based on an estimate of the delay distribution. This method is routinely used, for instance, to correct for reporting delays in AIDS cases. However it requires that all dates of onset or specimen collection are available. Furthermore it necessitates organism-specific calculations, since delays depend on identification procedures which may vary between organisms. A more fundamental objection is that outbreaks are imputed rather than observed, a procedure of questionable reliability.

The alternative is to analyse the data by date of receipt of report, which is necessarily always available. For comparability, the historical series should also be ordered by date of report. The major advantage of this method is that it is based on actual, rather than imputed data, although reporting delays inevitably affect detection sensitivity and specificity. However this may in itself have indirect benefits by shifting the focus of attention to improving the surveillance and reporting system, rather than tinkering with putative adjustments of dubious trustworthiness.

The *mean* of the reporting delay distribution affects the timeliness of outbreak detection, while its *variance* affects its sensitivity. In a surveillance system involving many reporting centres, it is therefore important to ensure that all centres work to the same protocol, with the same deadlines, in order to ensure that the variability of reporting delays are kept to a minimum. In addition, there should be a concerted attempt to reduce the average reporting delay concurrently in all centres. The mean reporting delay should ideally be small in relation to the time scale over which an outbreak might develop. For national outbreaks of food-borne infection, an outbreak might span several weeks: it is important therefore that delays between infection date and reporting date are kept within 1 or 2 weeks.

2.2 Systematic Variation

Typically, infectious diseases undergo large-scale systematic variations in incidence not attributable to outbreaks. These may be due to seasonal cycles, epidemic cycles, or secular trends. For instance, respiratory syncytial virus peaks in early winter, rotavirus infections peak in late winter/early spring, whereas most salmonellas peak in summer. Infections such as parvovirus B19 undergo regular epidemic cycles with epidemic periods of several years; influenza B in the UK on the other hand has an irregular epidemic cycle. Some infections are emerging as major new health problems, such as *Salmonella typhimurium* DT104 or verocytotoxin-producing *E. coli* O157 in the UK, which display increasing secular trends. Others, such as measles and other infections controlled by mass vaccination, are declining in importance.

Figure 2: Reports of Salmonella enteritidis PT4 England and Wales, 1990-1997

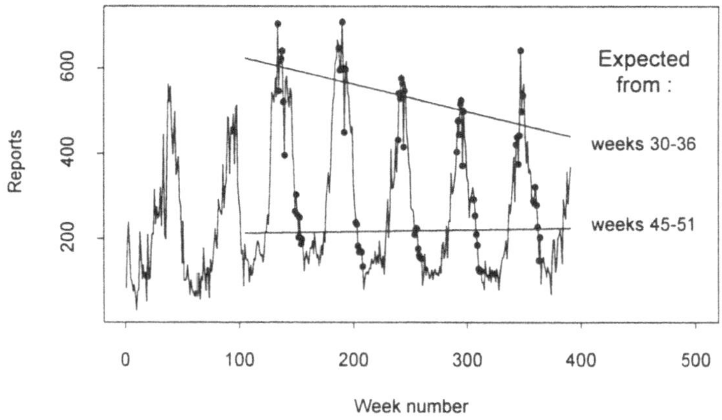

The major aim of an outbreak detection system is to detect sudden departures from regular patterns: generally, the aim is *not* to detect regular features, such as a seasonal increase in salmonellas. There are some specific exceptions to this: for instance, in the case of influenza, it may be important to know when an annual increase takes place, in order to heighten public awareness and reinforce control measures.

Regular fluctuations in incidence can easily be incorporated into statistical models. However some infections display complex patterns: for instance figure 2 shows that the secular trend in *Salmonella typhimurium* DT104 reports over the past five years varies according to the point in the epidemic cycle. In this case, attempts to model the entire historical series using independent trend and seasonal components will fail. More generally, this suggests that, for a given period, past data values not occurring at similar points in previous epidemic cycles are of little relevance in establishing baselines.

2.3 Influence of Past Outbreaks

In order to decide on what constitutes an outbreak-free level of infection counts, it is necessary to eliminate the impact of past outbreaks. In statistical terms, these correspond to outliers in the data series. Failing to adjust for such outliers can result in unrepresentative baselines, as shown in figure 3.

Figure 3: Reports of Salmonella Saint Paul
England and Wales, 1990-1997

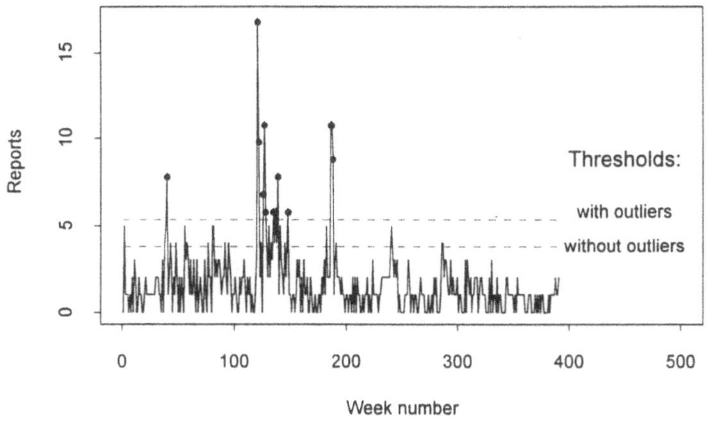

The easiest way to correct for past outbreaks is simply to leave them out of baseline calculations. However, while this is usually easily accomplished by eye, it is more difficult to devise an algorithm to recognise past outbreaks. Indeed, no entirely satisfactory solution has yet been found.

2.4 Robustness

The requirement of robustness stems from the need to process large numbers of distinct organisms rapidly and frequently. In the sense envisaged here, robustness is not required for outbreak detection systems constructed specifically to handle a single infection. These types of detection systems can incorporate the particular features of the organism of interest: we call such systems *organism-specific*. Organism-specific detection systems can thus use relatively complex modelling tools and exploit prior knowledge about the behaviour of that organism.

Figure 4: Numbers of organism types reported by week, England and Wales, 1996

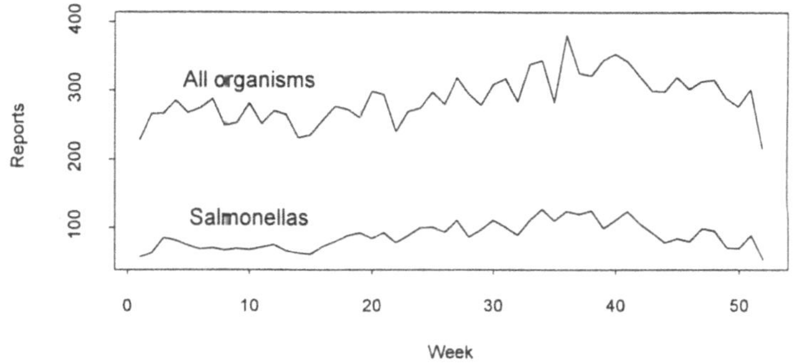

In contrast, in the surveillance of non-specific diseases such as gastro-enteritis, it is necessary to monitor the large numbers of different infective organisms which may be associated with clinical illness. It is usually required, for epidemiological purposes, to distinguish between subtypes. In the UK, for instance, Salmonellas are subdivided into serotypes and phage types. It is thus necessary to process large amounts of data relating to different organism types within a reasonable time-frame for investigation. Figure 4 shows the total numbers of different organisms, and of distinct Salmonella serotypes and phage types, reported to the Communicable Disease Surveillance Centre in 1996. There are in excess of 4,000 distinct organism types in the CDSC database. In order to process this volume of data each week within a matter of hours, relatively simple methods are required which do not rely on frequent re-setting of organism-specific parameters. For practical purposes, a single robust algorithm is required to handle the wide variety of organisms reported.

Robustness is required with respect to the following characteristics. Firstly, the system must be able to handle both rare and common organisms. The contrasts are striking: many rare salmonellas have weekly frequencies of less than 0.1, while others, such as *Salmonella enteritidis* PT4 run into hundreds of reports each week. Second, the algorithm must be able to cope with different seasonalities and trends, and with the absence of seasonality or trend. Third, it must have some mechanism to handle past outbreaks.

2.5 Sensitivity and false-detection rate

To be of any practical use, an outbreak detection system must be sensitive, while the false detection rate must be kept low. The false detection rate is defined as the proportion of apparently aberrant reports which are not due to outbreaks. A high false detection rate is likely to jeopardise confidence in the system.

Most outbreak detection systems are based upon a statistical criterion, such as a confidence or prediction interval. One of the shortcomings of all such mechanisms is their inability to differentiate between statistical significance and practical importance. Thus one or two isolations of a rare organism might constitute an unusual event in statistical terms, but, unless the organism is highly pathogenic, would not necessarily warrant further epidemiological investigations. In practice, the sensitivity of the system might be controlled primarily by the resources available for further investigations, rather than a requirement to detect every aberrant count.

Controlling the false detection rate is perhaps a more difficult problem, particularly when processing large numbers of organisms, or investigating large numbers of locations. Inevitably, in view of the large scale multiple testing involved, the number of false detections is likely to be high. A further requirement is to ensure that the false detection rate is broadly constant over the range of organism frequencies.

Evaluating an outbreak detection system is in itself a methodological challenge. Estimating the false detection rate is especially difficult, since lack of external evidence that an identified excess is attributable to an outbreak does not establish that it is not an outbreak. Simulation studies can be used to evaluate the system's performance under controlled conditions, but these are unlikely to reflect the vagaries of surveillance data. A different, qualitative approach is to evaluate the system's usefulness as perceived by the epidemiologists who use it.

3. Review of Statistical Methods

As noted in the introduction, there exists a large statistical literature on the detection of temporal and spatio-temporal clusters. Most of this work, however, has focused on the detection of retrospective clusters, that is, clusters which may have occurred in some past time interval. As noted above, this contrasts with the problem of detecting outbreaks prospectively, that is, as they arise. In addition to the methodological problems noted in the previous sections, prospective and retrospective detection differ fundamentally in the types of hypotheses to be tested. Thus in prospective detection, interest focuses on the null distribution of organism counts at the current time t, given the history of the process prior to time t. This requires the specification of a) an appropriate outbreak-free distribution, and b) a decision rule by which the current observed value is classified as aberrant or not. In contrast, most statistical procedures for retrospective cluster detection are designed to identify any clusters in some specified time interval (t_1, t_2) with $t_1 < t_2 < t$, where t is the current time. Some of these methods can be adapted for use in prospective detection. Others however, such as the scan statistic (Wallenstein 1980) or Tango's statistic (Tango 1984) are designed to test a non-specific hypothesis of no past clustering: these methods are perhaps less appropriate for prospective detection.

In this section we review some of the statistical methods which have been developed specifically for the purpose of prospective outbreak detection, or which have been used for retrospective purposes but which are likely also to be useful prospectively. Our aim is not to conduct an exhaustive review of the literature on cluster detection and its applications, but rather to illustrate the diversity of techniques available.

3.1 Regression methods

A conceptually simple, hence appealing, approach is to estimate the distribution of counts Y at time t, f(y|t), by means of a regression of the observed counts y_i, i=1,...,n at previous times t_i. For instance, if the counts are large, f(y|t) might be assumed $N(\mu, \sigma^2)$; for low counts, a Poisson distribution may be appropriate. The expected count μ is estimated using the predicted value at time t, and an upper threshold may be derived using prediction limits. An advantage of this method is that the regression equation can readily incorporate terms for trend and seasonality. A shortcoming is that it ignores any serial correlation which may exist between counts. Note however that, for rare organisms, serial correlation is not an issue since, at least in non-outbreak conditions, most cases are sporadic.

3.1.1 Example 1: the MMWR system

The simplest version of this method is used in the aberrant report detection system described by Stroup, Williamson & Hearndon (1989). This is a routine monitoring system applied to monthly counts of notifiable diseases in the United States. The output, published each month in the Morbidity and Mortality Weekly Report, published by the Centers for Disease Control and Prevention, displays a barchart of the observed over expected ratios, with 95% confidence limits.

This system does not incorporate time trends, and handles seasonality by comparing the current month with the same month and the two adjacent months in each of the 5 previous years. Thus the September 1997 count is compared to counts in August, September and October of 1992 to 1996. In this system the predicted value is the mean of the 15 baseline values. A 95% confidence interval for the current value is calculated using the sample mean and variance of the baseline observations, and the current observed over expected ratio is declared aberrant if it lies outside the corresponding confidence interval.

The considerable advantage of this system is its simplicity, and the ease with which the output can be conveyed to the epidemiological community. This approach, or a simple variation of it, is widely used in the detection of nosocomial outbreaks: see, for instance, Birnbaum (1984). The major shortcoming of the method is that it does not incorporate any correction for past outbreaks. These can seriously distort the baselines, resulting in an inflated mean, and hence an observed over expected ratio which is artificially low. In addition, the Normality assumption, which is valid for the substantial counts of notifiable diseases, is unlikely to hold for less common infections. Kafadar and Stroup (1992) discuss the impact of serial correlation on their method, and conclude that the simple estimation method described above is preferable to more complex resampling schemes, unless detailed information is available on the underlying correlation structure. An evaluation of the system is described in Stroup *et al.* (1993).

3.1.2 Example 2: Explicit modelling of trend and seasonality

In a pioneering application, Serfling (1963) used regression techniques to quantify the mortality attributable to influenza in cities in the United States. Although the focus of this application differs from that of outbreak detection, many of the ideas are directly transferable to this context: see, for instance, Costagliola *et al.* (1991) and Flahault *et al.* (1995).

Essentially, the model fitted is a Normal errors regression, with a linear secular time trend and sinusoidal seasonal components. Thus the expected value at time t is given by the regression equation:

$$E(y|t) = \alpha + \beta t + \gamma \cos(\omega t + \varphi)$$

Serfling fitted his model in three stages. First, he estimated the secular trend from selected periods of low incidence, mid-May to mid-September. He then estimated the seasonal parameters from the de-trended data for non-epidemic years, the epidemic years being identified by eye. Finally he estimated the (null) residual variance from the sum of squared residuals corresponding to non-epidemic years.

This procedure may easily be adapted to the construction of suitable thresholds for future periods, and hence may be applied to the prospective detection of temporal clusters. The expected value obtained by this method is adjusted for past epidemics, and incorporates adjustments for trend and seasonality. However the method involves a considerable degree of empirical judgement, particularly in identifying past epidemic years. It is thus not well suited for automation. Furthermore, an explicit functional specification of the seasonal component is unlikely to work when applied to a broad spectrum of organisms with different, and sometimes non-existent, seasonalities.

3.1.3 Example 3: Poisson regression

Both of the examples discussed above use Normal errors regression, which is perfectly suitable when counts are large, but inappropriate when dealing with infrequent events. Parker (1989) describes a cluster detection system to detect changes in abortion-related mortality in the United States, using Poison regression methods. The model is a log-linear model of the form:

$$Y_t \approx P(\mu_t) \quad , \quad \log(\mu_t) = \alpha + \beta t$$

where Y_t is the observed incidence rate of abortion-related deaths at time t. The prediction variance for an individual observation at time t is then given, approximately, by:

$$\hat{\mu}_t \, [1 + \hat{\mu}_t \, (\hat{\alpha} + 2t \operatorname{cov}(\hat{\alpha}, \hat{\beta}) + t^2 \hat{\beta})]$$

which may be used to construct a threshold value.

The advantage of this approach is that, unlike Normal errors regression, it is applicable to rare events. However, the Poisson assumption is unlikely to hold in many circumstances, and some method of handling extra-Poisson variation is required. A further requirement with the Poisson approach is to incorporate a skewness correction into the threshold calculation, so that the false-positive probability (that is, the probability of the current count exceeding the threshold when there is no outbreak) remains constant over a broad range of expected values. In particular, thresholds derived from symmetric prediction intervals are unsuitable.

3.1.4 Example 4: The CDSC system

The detection system in use at the Communicable Disease Surveillance Centre, described in Farrington *et al* (1996), is the only one to our knowledge which is specifically designed to handle a wide variety of organism frequencies and temporal patterns. The basic model is a quasi-likelihood log-linear regression model:

$$\log(\mu_i) = \alpha + \beta t_i \quad , \quad E(y_i) = \mu_i \quad , \quad V(y_i) = \varphi \mu_i$$

Seasonality is handled as described in section 3.1.2. In an attempt to reduce the influence of past outbreaks, baseline values with high residuals are given lower weights in the regression: for a baseline count with standardised residual s_i, the weight is γs_i^{-2} if $s_i > 1$ and equal to γ otherwise, for some constant γ.

In order to ensure a broadly constant false positive rate over the wide range of organism frequencies, the threshold is calculated after applying a 2/3 power transformation. The threshold value is the upper $100(1-\alpha)\%$ prediction limit:

$$U = \hat{\mu}_0 \left\{ 1 + \frac{2}{3} z_\alpha \left(\frac{\hat{\varphi}\,\hat{\mu}_0 + \mathrm{var}(\hat{\mu}_0)}{\hat{\mu}_0^{\,2}} \right)^{1/2} \right\}^{3/2}$$

The power 2/3 was chosen as it is the transformation to approximate symmetry for the Poisson distribution. The rationale for this model is that, for rare organisms, $\varphi \approx 1$ and are distributed Poisson, whereas for more frequent organisms Normal approximations are valid.

The model is run automatically every week on all 250-350 different organism types reported. Farrington *et al.* (1996) discuss the assumptions underlying the model and give an evaluation of its performance.

3.2 Time series methods

Time series methodology is an obvious candidate for detection systems. After all, the surveillance data upon which the detection system is based is time series data. Furthermore, time series methods are unique in that they exploit the correlation structure of the data. However, there are two major problems associated with time series methods in the present context. First, these methods generally require an initial transformation to achieve stationarity. Second, the model-fitting techniques required, at least for the more complex Box-Jenkins models, are difficult to automate, in that they require careful interpretation of a panoply of autocorrelation functions.

3.2.1 Example 5: Exponential smoothing

The application of exponential smoothing to monitoring surveillance data was first suggested by Healy (1983). The method has been applied by Ngo, Tager and Hadley (1996) to the detection of nosocomial outbreaks.

The exponential smoothing model is essentially a weighting procedure, with greater weight given to more recent observations. Thus the prediction at time t+1, $\hat{y}_{t+1}$, is related to the observed values by:

$$\begin{aligned}\hat{y}_{t+1} &= \alpha y_t + \alpha(1-\alpha)y_{t-1} + \alpha(1-\alpha)^2 y_{t-2} + \ldots \\ &= \alpha y_t + (1-\alpha)\hat{y}_t\end{aligned}$$

The forecast errors may then be used to estimate the forecast error variance, and hence to define threshold values.

In the nosocomial application envisaged by Ngo, Tager and Hadley, the forecast error was found not to depend sensitively on the choice of the smoothing constant α. However the method as described is only applicable to stationary series. In particular, recent outbreaks would seriously affect the accuracy of the forecasts.

3.2.2 Example 6: Box-Jenkins models

Watier, Richardson and Hubert (1991) describe the construction of a warning system for outbreaks of Salmonella infection, using Box-Jenkins time series models. They fit a seasonal autoregressive integrated moving average (SARIMA) model to data on *Salmonella Bovismorbificans*. The model may be written:

$$(y_y - y_{t-12}) = \alpha(y_{t-1} - y_{t-13}) + \varepsilon_t + \beta\varepsilon_{t-12}$$

where y_t is the count in period t. Prior to fitting the model, they identify past outbreaks and replace these counts with those expected had there been no outbreak. The alert threshold is calculated from the upper limit of a suitably chosen confidence interval on the 1-step forecast, with an adjustment for small fluctuations in seasonal effects.

Watier, Richardson and Hubert (1991) give careful consideration to the methodological problems inherent in devising outbreak detection systems. Their approach incorporates many adjustments to cater for these difficulties. However, while the authors note that their approach is applicable to other Salmonella serotypes, it is difficult to automate in such a way as to guarantee robustness for a wide variety of organisms.

Choi and Thacker (1981) also described a time series approach, for detecting outbreaks of influenza. Interestingly, they compared the performance of this method to that of Serfling (see section 3.1.3), and concluded that the time series method is superior. However they also note that if, in the regression method, counts in epidemic years are replaced by expected values rather than being excluded, then the two methods yield comparable results.

3.3 Statistical process control methods

The problem of prospectively detecting outbreaks of infectious disease bears some similarity to that of detecting aberrances in industrial production processes. It is no surprise therefore that several approaches have been proposed which are directly inspired by, or related to, methods of statistical process control. For instance many of the methods used for nosocomial surveillance are similar to Shewart charts (Birnbaum 1984, Parkhurst *et al.* 1985). In this section we consider three methods which share a flavour of statistical process control methodology. Other such methods have been proposed, for instance the short memory scheme of Shore and Quade (1988) based on the distribution of cases in the current and previous periods, but will not be considered further here.

3.3.1 Example 7: Cumulative sum (CUSUM) techniques

Cumulative sums (cusums) were introduced by Page (1954) and have been used, for instance, for detecting the onset of influenza epidemics in the UK (Tillett and Spencer 1982) and salmonella outbreaks in the USA (Hutwagner *et al.* 1997). Given counts y_1, ..., y_t with expected values $\mu_1,\ldots,\mu_t$ and common variance, the cusum at time t is:

$$z_t = \sum_{i=1}^{t}(y_i - \mu_i) .$$

In the absence of any systematic departure from the expected values μ_i, the cusum varies stochastically around zero. However if after some time t_0 the mean increases by some value α, then the cusum will have expectation $\alpha\ (t - t_0)$: thus the cusum will register a change in slope.

Formal statistical methods for analysing cusum charts are described in standard textbooks on statistical process control (see for instance Mittag and Rinne 1993). Cusums are sensitive to small changes in numbers of reports, and hence are well suited for detecting the timing of onsets of expected epidemics such as influenza. However, for the same reason, they are also sensitive to small changes in reporting efficiency and other artefacts of the reporting process, and hence may lack robustness when used with surveillance data unless the baselines are frequently reset.

3.3.2 Example 8: Methods based on times to failure

When events are very rare, methods based on the monitoring of aggregate total cases in discrete time periods will fail, since even one event is 'unusual' in an statistical sense. One way round this problem is to specify a minimum outbreak size. Another approach is to base detections not on total reports, but on the intervals between them. This is the approach developed by Chen (1978) for tracking congenital abnormalities.

Assuming that events arise in a stationary Poisson process, inter-event times follow an exponential distribution with mean μ, say. The detection threshold is specified by parameters (n, τ): an aberrance is declared if the intervals between $n+1$ consecutive events are all less than τ. The probability of this occurring if there is no change in the underlying rate, that is, the false detection probability, is $(1-e^{-\tau/\mu})^n$. Similarly, the sensitivity of detection when the rate of events increases by a factor γ is $(1-e^{-\gamma\tau/\mu})^n$.

The background rate, and hence μ, may be determined from historical data. Appropriate values of n and τ are selected so as to yield acceptable sensitivity and false detection probabilities. The analysis is undertaken whenever an event occurs. Chen, Connelly and Mantel (1993) propose supplementing the method with a further confirmatory procedure, in order to reduce the false detection probability.

This method overcomes some of the shortcomings of other approaches when events are rare, for example with very uncommon Salmonella serotypes, and might be useful to detect low-level point-source outbreaks. Barbujani (1987) reviews this and other methods for monitoring malformations.

3.3.3 Example 9: Derivative-based methods

All the methods previously described other than the last are based on the calculation of a threshold value, above which reports are declared aberrant. An intuitively appealing alternative suggested by Nobre and Stroup (1994) is instead to base detections on changes in the first derivative of the data series with respect to time. Two potential advantages of this approach are that departures from the norm might be detected earlier, before reports build up to the threshold value, and that long series of historical data are not required.

For simplicity, consider a simple exponentially smoothed model (see section 3.2.2 above). The forecast error $e_t = y_t - \hat{y}_t$ may be written:

$$e_t \quad = \quad y_t^1 + (1-\alpha)\, y_{t-1}^1 + (1-\alpha)^2\, y_{t-2}^1 + \ldots$$

where $y_t^1 = y_t - y_{t-1}$ is the first numerical derivative. The forecast error is thus a weighted sum of the previous values of the numerical derivative. In the absence of systematic deviations from stationarity, the forecast error will remain close to zero. Nobre and Stroup (1994) further smooth the forecast error, and convert it into an index function which varies between -1 and +1. Departures of the index function beyond suitable limits then indicate aberrant behaviour.

Nobre and Stroup (1994) extend this approach to data series which can be modelled by polynomial functions, and discuss an application to data on measles in the USA. The main shortcoming of the approach is its lack of specificity in some circumstances, in that small fluctuations in the data can produce large derivatives. The authors propose to remedy this by introducing further constraints on the detection criterion.

3.4 Methods incorporating spatial information

All the methods described so far are based on detecting changes in the *temporal* pattern of reports, using data which have been aggregated over geographical areas. National detection systems are unlikely to be more effective in detecting *local* outbreaks than more traditional surveillance methods by epidemiologists and other staff, based on experience of local conditions. However in some circumstances it might be possible to enhance national (or international) systems by incorporating information on *spatial* heterogeneity. While there is a large literature on the detection of space-time interactions (see, for instance, Knox 1964 and Mantel 1967), very little has been written on corresponding methods for outbreak detection. We tentatively discuss three possible areas of application.

3.4.1 Example 10: Spatio-temporal detection criteria

A simple approach to incorporating spatial information in detection criteria is to calculate location-specific thresholds, and to declare an alert if the threshold is exceeded in a specified number of places. For progressive epidemics this may allow earlier detection, since it is not required that the aggregate total reports exceeds an overall threshold.

Furthermore, in some circumstances, it makes little sense to aggregate data over geographical units. This holds for some multi-national surveillance systems in which participant countries use different methods of data collection, with different completeness of reporting. In such circumstances it is essential to undertake an analysis within countries.

One example of such a system is Salm-Net (Fisher *et al.* 1994), a surveillance network which collects data on Salmonella infections from 14 European countries. The data from each individual country is analysed using the algorithm described in section 3.1.5. A multi-national outbreak is declared if two or more countries register totals above their respective threshold values.

3.4.2 Example 11: Smoothing methods

One of the obstacles to developing detection systems with a spatial dimension is the problem of controlling the number of false detections. As the number of geographic units increases, so does the probability of a spurious detection. One possible solution is to smooth the data by incorporating information from neighbouring areas. Raubertas (1989) has proposed a spatial version of the cusum statistic, incorporating information on geographical proximity. Suppose that data are available from k locations, with some proximity matrix **D**, where d_{ij} = proximity of locations i and j. Let $\mathbf{y}_t$ denote the vector $(y_{1,t},\ldots,y_{k,t})$ of reports at time t from each of the k locations, with baseline expectation μ_t. Then for each location i, let:

$$\tilde{z}_{it} = \sum_{j=1}^{k} d_{ij} \sum_{s=1}^{t} (y_{js} - \mu_{js})$$

This cusum is a weighted sum of the individual cusums for the data from each location. Smoothing the data in this way reduces the influence of individual locations. The choice of proximity matrix will depend on the infection concerned, and could, for instance, be non-zero only for surrounding locations.

3.4.2 Example 12: Epidemic mapping

Smoothing techniques can be further extended to provide dynamic epidemic maps which chart the spatio-temporal progression of epidemics. For instance, Carrat and Valleron (1992) apply kriging methods to smooth the spatial distribution of reports in each period. Such techniques, combined with a threshold criterion to indicate an unusual excess of cases, could usefully enhance outbreak detection mechanisms by providing more detailed spatial information. It may also be possible to develop them further so as to provide information on the direction and velocity of spread.

4. Conclusion

The test of an outbreak detection system is that it is perceived as *useful* by epidemiologists and other professionals involved in public health medicine. This requires that the system detect outbreaks, and that epidemiologists recognise its value.

Unfortunately, the track record of computer-based, statistical methods of outbreak detection has so far been a mixed one. Although widely used in hospital settings to detect nosocomial outbreaks, one prominent writer on the topic concluded:

> "I now believe that computers are required for applied research, may be invaluable for certain descriptive epidemiology, but are not essential for routine infection surveillance" . (Birnbaum 1987)

Similarly, for community outbreaks, it is undeniable that the most effective outbreak detection mechanism is a network of epidemiologists and microbiologists who, through experience and knowledge of the community which they serve, can rapidly recognise unusual increases in infections, without having to resort to more complex statistical methods.

Nevertheless, we argue that computer-based outbreak detection systems do have a role to play in complementing other more traditional methods, though they cannot and should not aspire to replace them. First, an automated computer detection system can act as a 'safety net' which routinely scans reports, and thus can identify any problems which, for whatever reason, may have been overlooked. This is particularly important when data on large numbers of organisms are collected, which it is not physically possible to process manually. Second, national outbreaks may not be detected locally if they do not result in substantial local increases in numbers of cases. A third, indirect, benefit is that implementing computer-based procedures can help to motivate improvements in the collection of surveillance data.

An essential ingredient for a successful computer-based detection system is close co-operation between epidemiologists, microbiologists, computer scientists and statisticians. To maximise its effectiveness, outbreak detection should not rely on a single system, but ideally should seek to combine different methods, including computer-based screening procedures. Such an inter-disciplinary, collaborative approach can enhance existing methods, and improve the data collected in the process.

References

Barbujani G. (1987). A Review of Statistical Methods for Continuous Monitoring of Malformation Frequencies. *European Journal of Epidemiology*, **3**, 67-77.

Birnbaum D. (1984). Analysis of Hospital Surveillance Data. *Infection Control*, **5**, 332-338.

Birnbaum D. (1987). Nosocomial Infection Surveillance Programs. *Infection Control*, **8**, 474-479.

Carrat F. and Valleron A.-J. (1992). Epidemiologic Mapping Using the "Kriging" Method: Application to an Influenza-like Illness Epidemic in France. *American Journal of Epidemiology*, **135**, 1293-1300.

Chen R. (1978). A Surveillance System for Congenital Malformations. *Journal of the American Statistical Association*, **73**, 323-327.

Chen R., Connelly R.R., and Mantel N. (1993). Analysing Post-Alarm Data in a Monitoring System in Order to Accept or Reject the Alarm. *Statistics in Medicine*, **12**, 1807-1812.

Costagliola D., Flahault A., Galinec D., Garnerin P., Menares J. and Valleron A.-J. (1991). A Routine Tool for Detection and Assessment of Epidemics of Influenza-like Syndromes in France. *American Journal of Public Health*, **81**, 97-99.

Farrington C.P., Andrews N.J., Beale A.D. and Catchpole M.A. (1996). A Statistical Algorithm for the Early Detection of Outbreaks of Infectious Disease. *Journal of the Royal Statistical Society*, Series A, **159**, 547-563.

Fisher I.S.T., Rowe B., Bartlett C.L.R. and Gill O.N. (1994). Salm-Net: Laboratory-Based Surveillance of Human Salmonella Infections in Europe. *PHLS Microbiology Digest*, **11**, 181-182.

Flahault A., Garnerin P., Chauvin P., Farran N., Saidi Y., Diaz C. and Toubiana J. (1995). Sentinelle Traces of an Epidemic of Acute Gastroenteritis in France. *Lancet*, **346**, 162-163.

Healy M.J.R. (1983). A Simple Method for Monitoring Routine Statistics. *The Statistician*, **32**, 347-349.

Hutwagner L.C., Maloney E.K., Bean N.H., Slutsker L. and Martin S.M. (1997). Using Laboratory-Based Surveillance Data for Prevention: An Algorithm for Detecting *Salmonella* Outbreaks. *Emerging Infectious Diseases*, **3**, 395-400.

Kafadar K. and Stroup D.F. (1992). Analysis of Aberrations in Public Health Surveillance Data: Estimating Variances on Correlated Samples. *Statistics in Medicine*, **11**, 1551-1568.

Knox E.G. (1965). The detection of space-time interactions. *Applied Statistics*, **13**, 25-29.

Mantel N. (1967). The detection of disease clustering and a generalised regression approach. *Cancer Research*, **27**, 209-220.

Mittag H.-J. and Rinne H. (1993). *Statistical Methods of Quality Assurance*. Chapman and Hall, London.

Ngo L., Tager I.B. and Hadley D. (1996). Application of Exponential Smoothing for Nosocomial Infection Surveillance. *American Journal of Epidemiology*, **143**, 637-647.

Nobre F.F. and Stroup D. (1994). A Monitoring System to Detect Changes in Public Health Surveillance Data. *International Journal of Epidemiology*, **23**, 408-418.

Page E.S. (1954). Continuous Inspection Schemes. *Biometrika*, **41**, 100-115.

Parker R.A. (1989). Analysis of Surveillance Data with Poisson Regression: A Case Study. *Statistics in Medicine*, **8**, 285-294.

Parkhurst S.M., Blaser M.J., Laxson L.B. and Wang W.-L. L. (1985). Surveillance for the Detection of Nosocomial Infections and the Potential for Nosocomial Outbreaks. *American Journal of Infection Control*, **13**, 7-15.

Raubertas R.F. (1989). An Analysis of Disease Surveillance Data that Uses the Geographic Locations of the Reporting Units. *Statistics in Medicine*, **8**, 267-271.

Serfling R.E. (1963). Methods for Current Statistical Analysis of Excess Pneumonia-Influenza Deaths. *Public Health Reports*, **78**, 494-506.

Shore D.L. and Quade D. (1989). A Surveillance System Based on a Short Memory Scheme. *Statistics in Medicine*, **8**, 311-322.

Stroup D.F., Williamson G.D. and Herndon J.L. (1989). Detection of Aberrations in the Occurrence of Notifiable Diseases Surveillance Data. *Statistics in Medicine*, **8**, 323-329.

Stroup D.F., Wharton M., Kafadar K. and Dean A.G. (1993). Evaluation of a Method for Detecting Aberrations in Public Health Surveillance Data. *American Journal of Epidemiology*, **137**, 373-380.

Tango T. (1984). The Detection of Disease Clustering in Time. *Biometrics*, **40**, 15-26.

Tillett H.E. and Spencer, I.-L. (1982). Influenza Surveillance in England and Wales using Routine Statistics. *Journal of Hygiene Cambridge*, **88**, 83-94.

Wallenstein S. (1980). A Test for Detection of Clustering over Time. *American Journal of Epidemiology*, **111**, 367-372.

Watier L., Richardson S. and Hubert B. (1991). A Time Series Construction of an Alert Threshold with Application to *S. Bovismorbificans* in France. *Statistics in Medicine*, **10**, 1493-1509.

REPRESENTING AGGREGATION AND SCALE EFFECTS IN HIV/AIDS EPIDEMIC MODELLING SYSTEMS

Richard Thomas
School of Geography
University of Manchester
Oxford Road
MANCHESTER M13 9PL, UK
email: r.thomas@man.ac.uk

Abstract
This paper reviews and evaluates some of the alternative disaggregations of the population at risk to infection that have been specified in HIV/AIDS epidemic modelling systems, together with methods for simulating incidence in space and time. Particular attention is paid to the influence of scale (the nature and size of the geographical units analysed) as a constraint upon both the model designs and their application to understanding the consequences of disease prevention strategies.

1 Introduction

In a discussion of approaches to modelling and predicting the incidence of HIV/AIDS, Solomon (1996) makes the distinction between short and long term forecasting methods. The former are concerned with deriving statistical fits for observed AIDS incidence to enable the prediction of current HIV prevalence and account for reporting delays. Popular procedures include extrapolative forecasting (Rosenberg, 1992), empirical Bayes estimates (Mariotto, 1989), backcalculation (Raab et al, 1994) and multistage semi-Markov models for representing the progression of those with HIV through the incubation period (Longini et al, 1992). This paper, however, is concerned with the more extended perspective that is afforded by mathematical models for the transmission of HIV infection which assist understanding qualitative aspects of the dynamics of HIV diffusion and the impact of prevention and intervention strategies on this progression (Smyth and Thomas, 1996a; 1996b; Thomas, 1996a; 1996b).

The initial developments of these modelling systems are documented in Anderson (1988) and Isham (1988). Their central concerns were the mathematical representation of the host-HIV relationship and the prediction of the size and timing of AIDS incidence amongst those thought to be most at risk to infection. The gradual evolution of the pandemic, however, has seen other issues emerge on the modelling agenda. In a recent summary of the *Problem Areas* identified at a NATO workshop on Epidemic Models, Mollison (1995) stressed the need to take more direct account of spatial effects on the transmission process. The simpler models, for example, usually assume the population at risk mix homogeneously, yet geography introduces many demographic heterogeneities which serve to buffer the diffusion of infection. This paper, therefore, reviews and evaluates some of the alternative demographic disaggregations that have been proposed for modelling HIV/AIDS together with methods for simulating incidence in space and time. Particular attention is paid to the role of scale (the nature and size of the geographical units analysed) as a constraint upon both model designs and their application to understanding the consequences of preventative actions.

2. Disaggregations for Infection Risk

2.1 Behavioural Strategies

Models for simulating HIV infection and AIDS incidence are adaptations of standard systems for the transmission of an infectious disease agent among those at risk (Bailey, 1975; Haggett, 1994). The simplest approach to this task is to develop models for a single risk behaviour (homosexuality, heterosexuality, intravenous drug injection) and, thereby, presume the degree of mixing between these populations may be regarded as negligible. The models of Anderson (1988) and Isham (1988), for example, are applicable to a single community of gay men. Suppose this community numbers n individuals and x denotes the number who are susceptible and y those who are infectious. Then, at any point in time, the transmission of HIV is represented by the following state diagram.

$$\mu n \rightarrow [x] \rightarrow \beta rxy/n \rightarrow [y] \rightarrow y/D \qquad (1)$$

Here, μ is the rate of entry of men into the gay population and the expression $\beta rxy/n$ measures the incidence of HIV. Within this term, r is the number of sexual partners per unit of time, y/n is the probability a new partner is infectious and β is the probability an infectious partnership results in the transmission of HIV. The removal of infectives from active circulation depends upon D (the average duration of the period of communicability for HIV), such that y/D counts the number removed in each interval of time.

It is now widely recognised (Anderson and May, 1991; Heesterbeek, 1992; Mollison, 1995) that a key property of this system for interpreting disease control may be derived as an approximation from the simulated start of an epidemic when a single infection arises in an otherwise wholly susceptible population. This idea is expressed by the reproduction rate (R_0) of HIV infection given by

$$R_0 = \beta rD. \qquad (2)$$

Here, R_0 is seen to be the average number of susceptibles infected by the index case during the period of communicability. Consequently, $R_0 > 1$ (the index case more than reproduces himself prior to removal) is the threshold condition for an epidemic to begin.

This model for single behaviour, however, takes no account of the interactions that are facilitated by intermediate risks like bisexuality, prostitution and the sexual behaviours of intravenous drug users (IVDUs). This task requires a compartmental design where each risk behaviour is specified separately and the HIV incidence terms contain cross infection mechanisms for the feasible group interactions. Knox (1986; 1993) was one of the first to develop a model of this type. Unfortunately, the absence of precise data for many of the parameters and populations forced him to adopt an approximate calibration procedure which, in hindsight, has been found to have overestimated the extent and current size of the epidemic in the UK. Moreover, the complex designs of many compartmental models (Van Druten et al, 1990) have led some to argue that, beyond a certain refinement threshold, they add little to the predictive capabilities of less elaborate systems (Kaplan, 1989; Kaplan and Lee, 1989; Brookmeyer, 1991). It has also been suggested that such preoccupation with fine structure is counter productive for the extensive geographical units which usually provide the jurisdiction for health decision-making and forward planning (Bailey, 1995).

2.2 Activity Related Specifications

A major limitation of the models described so far is their adoption of an *average* partnership rate, r, which presumes each member of the community has the same frequency of risk activity. However, observations on samples of gay men during the early phases of the epidemic in San Francisco found behaviours ranging from celibacy to extreme promiscuity (Grant et al, 1987). This type of frequency distribution is similar to those observed for other STDs, where small core groups of individuals who acquire numerous partners play a disproportionate role in the maintenance and transmission of infection (Yorke et al, 1978). To imitate the outcome of this effect, Anderson and May (1991) have derived an extension of model (1) where the state variables are subscripted according to $r = 0, 1, . .$ to denote the number of people in each state who acquire *exactly* r partners per unit of time. They assume the rate of infection for a susceptible with r partners is linearly proportional to the average number of partners per unit of time, such that the probability a new partner of a susceptible from r is infectious, p_r, is

$$p_r = \Sigma_r r y_r [\Sigma_r r n_r]^{-1}, \quad \forall r. \tag{3}$$

Then, the state diagram for their specification is given by

$$\mu n_r \rightarrow [x_r] \rightarrow \beta r x_r p_r \rightarrow [y_r] \rightarrow y_r/D, \quad \forall r. \tag{4}$$

In comparison to model (1), simulations from this revised specification have shown that the activities of those with frequent partner changes raise the prevalence of HIV during the early phases of the epidemic. However, because these individuals are also removed from circulation with AIDS relatively early in the epidemic, they play little part in the later phases when transmission relies increasingly on the less sexually active and when HIV prevalence is less than that predicted by model (1). Consequently, the decline in the rate of seroconversion among gay men in San Francisco observed in the mid-1980s might have been more the result of the early onset of AIDS among those with most partners than an effect of the preventative actions taken by this community (Thomas and Smyth, 1997).

In essence, this particular disaggregation adds detail to the representation of the epidemic for a single risk behaviour. A more generalised version of this framework, however, is to model the inter-relationships between one or more core groups and the remainder of the population (Hethcote and Yorke, 1984; Thomas, 1994; Jacquez et al, 1995). Accordingly, assume the population is partitioned into $l = 1, . . , L$ groups each with a characteristic *average* partnership rate, r_l. Let q_{lk} denote the fraction of group l's contacts that are made with group k; $\Sigma_k q_{lk} = 1$. Then, the probability that the partner of a susceptible from l is infectious, p_l, is

$$p_l = \Sigma_k q_{lk} y_k / n_k, \quad \forall l, \tag{5}$$

and the corresponding state diagram is given by

$$\mu n_l \rightarrow [x_l] \rightarrow \beta r_l x_l p_l \rightarrow [y_l] \rightarrow y_l/D, \quad \forall l. \tag{6}$$

To be realistic, the sexual mixing imitated by the incidence term should obey the conservation law that requires the number of contacts group l has with group k to equal the number of contacts k has with l. Therefore, the values calibrated for the partnership rates and contact fractions should satisfy the following symmetry conditions

$$r_l q_{lk} n_l = r_k q_{kl} n_k, \quad \forall lk. \tag{7}$$

Moreover, the mixing of the groups is likely to be influenced by their social, cultural and demographic composition and, in recent years, a considerable amount of research has been devoted to devising appropriate specifications for these characteristics (Haraldsdottir et al, 1992; Lin et al, 1993; Hethcote, 1996). Consider, the case where the groups only mix with their own kind which, for example, might occur with an exclusively heterosexual-homosexual categorisation. This limit is termed assortive (A) mixing and is defined by the contact fractions

$$q^A{}_{lk} = \delta_{lk}, \quad \forall lk, \tag{8}$$

where $\delta_{lk} = 1$ if $l = k$ and is 0 otherwise. The opposite bound is variously termed proportionate, disassortive, or randomised (R) mixing and describes the case where the number of contacts between the groups is proportional to the products of their activity rates and sizes; that is

$$q^R{}_{lk} = r_k n_k/C, \quad \forall lk, \tag{9}$$

where $C = \sum_k r_k n_k$ - the total number of contacts. This scheme is usually appropriate for heterosexual mixing where the contacts made by males are in proportion to those received from females. Finally, preferred mixing arises as a compromise between the two limits and applies to the more restricted frequency of sexual contact that might be expected between different age cohorts or ethnic populations.

These specifications have important implications for the simulation of infection. Morris (1991; 1995; 1996), for example, has demonstrated that for stable attributes with assortive tendencies, like race and sexual preference, infection may often remain isolated among those with high activity rates. In contrast, more diasassortive attributes, like age, enhance the opportunity for infection between cohorts. Similarly, Jacquez et al (1991; 1995) relate these processes to the reproduction rates which underscore the chosen mixing specification. They define

$$R_{0lk} = \beta r_k q_{kl} D, \quad \forall lk, \tag{10}$$

to be the number of infections in l caused by an index case in k during the incubation period. Then, making the summation

$$R_{0k} = \sum_l R_{0lk}, \quad \forall k, \tag{11}$$

they obtain the reproduction number for *all* the contacts of the index case in k. These rates were analysed to define the set of groups that form a core capable of initiating sustained infection amongst the complementary set. They demonstrate that $R_{0ll} > 1$ is sufficient for a local epidemic but may be inadequate for l to act as a core (initiator) group, which requires both $R_{0l} > 1$ and $R_0 > 1$.

A relatively simple application of this modelling framework to simulating the possible effects of core group activity is described in Thomas (1994), where the specification divides the community into $L = 2$ classes. Group 1 is those at low risk where the membership criterion is that the number of partners acquired by each individual be less than, or equal to, the critical partnership rate, r^*, which defines a global reproduction rate of unity (see equation 2); that is

$$r^* = [\beta D]^{-1}. \qquad (12)$$

Conversely, the frequency of partner acquisition in $l = 2$, the core, is always in excess of this critical rate and, therefore, a simulated epidemic can only begin among this group. The calibration is based upon some fragmentary data sources for the UK (Knox et al, 1993) where group 1 is presumed to comprise 90% of the sexually active population who acquire partners at $r_1 = 1.11$ per year, while the respective values for the core are 10% and $r_2 = 10$. The contact fractions between these populations are taken to be $q_{12} = q_{21} = 0.05$, which satisfy condition (7) and are indicative of a high degree of preferred mixing, while the epidemiological parameters were given published estimates of $\beta = 0.1$ and $D = 2$ years (May et al, 1989) to define a critical rate of $r^* = 5$ partners per year.

Simulations made on this setting revealed that a small incidence among group 1 was not self-sustaining and, instead, was entirely dependent on cross-infections from the epidemic among the core. This outcome suggested that the activity rates in group 1 were insufficient to maintain chains of infection amongst these individuals which is consistent with the observed incidence of HIV in most developed countries (Kirp and Bayer, 1992), where occurrences of the direct transmission between low risk heterosexual partners have been rare. Moreover, the simulations revealed the minor incidence in $l = 1$ to be contained entirely within the duration of the core epidemic cycle. Together, these dependencies suggest the continued modification of the frequency of risk activity amongst the core would eventually *switch off* the interactions that sustain the group 1 incidence and highlight the importance of targeting interventions at the core.

3 Geographical Considerations

3.1 Systems of Regions

The state diagrams for the previously described models define differential equations which control temporal change in the epidemic variables. Accordingly, these derivations all assume that simulated contacts are made within an infinitesimally small unit of space and time and, therefore, take no account of how the distance separating individuals might deter the likelihood of their meeting (Gould, 1993). For relatively small areal units of data collection, this effect might well be averaged into the estimated contact rates and, consequently, have little impact on the simulated incidence. However, if the intention is to make forecasts for more extensive systems of regions, like cities or nations, then model specifications should take explicit account of the heterogeneities introduced by their degree of separation.

The first space-time models were developed to forecast the diffusion of influenza among the cities of the former Soviet Union (Baroyan and Rvachev, 1967) and, subsequently, these systems have been refined and applied to predicting pandemics of both influenza and HIV/AIDS (Bailey and Estriecher, 1987). The basic specification of these models requires the study area to be partitioned into $i = 1, \ldots, m$ regions and the epidemic variables to be subscripted accordingly (x_i, y_i, $\forall i$). Let η_{ij} denote the observed rate of migration per person per unit of time from region i to region j and, $M_i(x)$ and $M_i(y)$, $\forall i$ be the net balance of migration in each region for a particular epidemic state. Then, these net migrations for infectives are calculated from

$$M_i(y) = \Sigma_j (\eta_{ji}y_j - \eta_{ij}y_i), \quad \forall i \tag{13}$$

and are incorporated in the following state diagram:

$$\begin{array}{ccccccccc} \mu n_i & \rightarrow & [x_i] & \rightarrow & \beta r x_i y_i / n_i & \rightarrow & [y_i] & \rightarrow & y_i/D, \quad \forall i. \\ & & \uparrow & & & & \uparrow & & \\ & & M_i(x) & & & & M_i(y) & & \end{array} \tag{14}$$

Simulations taken on this model show that regions with the highest rates of inmigration are predicted epidemic series which, eventually, will lead those of other regions. Moreover, the migration matrix causes a redistribution of the regional populations (n_i, $\forall i$) during the simulation which stabilises at their demographic equilibria (Thomas, 1992).

Migration rates tend to reflect relatively permanent and long-term relocations of people and, therefore, their purpose in model (14) is to serve as proxies for the more transient journeys and contacts that more frequently result in the transfer of infection between regions. For this reason, I have argued (Thomas, 1988) that a more natural approach to simulating the diffusion of an infectious disease agent is to respecify the contact mechanisms in the incidence term in accordance with the tenets of spatial interaction theory (Wilson and Bennett, 1985). These statistical procedures for estimating flows of people between regions require the regional populations to mix in a manner that is consistent with a negative exponential function of the form $\exp(-\lambda d_{ij})$. Here, d_{ij} is the distance between the centres of regions i and j, and λ is a parameter calibrated in the interval $[0, \infty]$ to represent the degree to which distance curtails contact frequencies in the chosen study area. When $\lambda = 0$, all contacts occur irrespective of distance and, conversely, $\lambda = \infty$ makes all contact between regions impossible. Then, the probability a partner of a susceptible in region i is infectious, p_i, is

$$p_i = [\Sigma_j y_j \exp(-\lambda d_{ij})]/W_i, \quad \forall i, \tag{15}$$

where the W_i, $\forall i$ are given by

$$W_i = \Sigma_j n_j \exp(-\lambda d_{ij}), \quad \forall i. \tag{16}$$

These terms are proportional to adjusted regional populations for their accessibility to all others (Thomas, 1994). The state diagram for this specification is given by

$$\mu n_i \rightarrow [x_i] \rightarrow \beta r x_i p_i \rightarrow [y_i] \rightarrow y_i/D, \quad \forall i. \qquad (17)$$

The global reproduction rate for this model is given by equation (2), however, the rate in *any* region i may vary around the global rate depending on its relative accessibility to other regions (see Thomas, 1997). Moreover, in this representation the epidemiological parameters $\{\beta, r, D\}$ are specified as constants for all the regional populations. Accordingly, these parameters determine a constant duration for the epidemic cycle of HIV prevalence (y_i) *within* each region and act independently of the decay parameter, λ, which combines with distance to control the time elapsed *between* first infection in one region and another. In contrast to the long-term lags and leads simulated by the migration model, the regional series obtained from the spatial interaction format gradually synchronise with the evolution of the epidemic cycles (Thomas, 1990).

A further disaggregation of this system is to include the core group framework (model 6) within the design, such that n_{il} denotes the population size of group l in region i. Then, the equivalent accessibility adjustments are calculated from

$$W_{il} = \Sigma_j n_{jl}\exp(-\lambda d_{ij}), \quad \forall il, \qquad (18)$$

and the probability a partner of a susceptible from group l in region i is infectious, p_{il}, is

$$p_{il} = [\Sigma_k q_{lk}\Sigma_j y_{jk}(-\lambda d_{ij})]/W_{il}, \quad \forall il. \qquad (19)$$

The corresponding state diagram is given by

$$\mu n_{il} \rightarrow [x_{il}] \rightarrow \beta r_l x_{il} p_{il} \rightarrow [y_{il}] \rightarrow y_{il}/D, \quad \forall il, \qquad (20)$$

which imitates how an epidemic started in a particular group and region might be transferred to other cohorts and locations.

3.2 Applicability

One application of these multiregion modelling systems is to forecast the pathway of transmission ahead of a first infection observed in a particular source region. Longini et al (1986), for example, applied a version of the migration model (14), which employed airline passenger flow rates between $m = 52$ of the world's major cities as surrogates for migration terms, to simulate an accurate replication of the influenza pandemic that originated in Hong Kong in 1968 and persisted into 1969, while Flahault and Valleron (1992) have assessed the global diffusion of HIV based on air travel statistics. In a similar vein, I have obtained a close fit to the pathway of the rapid 1957 Singapore influenza pandemic using the spatial interaction model (17) with a distance matrix constructed for $m = 16$ major world cities (Thomas, 1993). While most deterministic epidemic modelling systems of this type are known to overestimate the time elapsed to epidemic peaks (Cliff et al, 1986; Cliff and Logan, 1997), these results have been sufficiently encouraging to promote further applications. The forecast time intervals from the source infection to first incidence in each other region, for example, serve as warning times for mounting immunisation campaigns and other interventions against the agent in future pandemics. Moreover, by simulating pandemics started in each region, their potency as sources may be assessed by the time taken for each seeded infection to traverse the globe (Thomas, 1993).

A similarly applicable outcome, however, does not apply to the prediction of the pandemic pathway of HIV/AIDS (Thomas, 1994; 1996a). Here, simulations drawn on the spatial interaction framework presumed the first human infection with HIV occurred in Zaire in 1959 (Gotlieb et al, 1981). It was found that a value of $\lambda = 2$ best fitted the early timing of the serologically dated pathway from strains of HIV-1 (Li et al, 1988) which indicates that, after its African origins, the virus diffused through the Caribbean to the US by the mid-1970s and then to Europe by about 1980. Unfortunately, this model also predicted a full scale epidemic in Asia well before this last date which is in contradiction with serological evidence that has adjudged HIV to have been in circulation in India only since 1982 (Banerjee, 1989). Moreover, the forecast indicated that, according to the relationship between the rate of diffusion and place accessibility which underpins spatial interaction theory, the most likely natural pathways out of Africa are through the Middle East and the Mediterranean, and not the Caribbean. One explanation for this divergence might be that the serological pathway is subject to bias because the early geographical concentration of HIV research on patients in Africa and the US overlooked other contingencies. Alternatively, the predominantly sexual partnership networks, which facilitated the transfer of infection to the US, might have been of sufficient intensity to break the constraints of place accessibility. These outcomes might be diminished if model (17) was reconstructed either with region-specific distance decay parameters, λ_i, $\forall i$, to imitate variable contact frequency with distance for each region, or with, λ_{ij}, $\forall ij$, for the equivalent effect on each route.

Even with such modifications, it seems unlikely that the ensuing warning times for HIV would be as applicable for disease control as those derived for influenza. In this respect, it is important to note that HIV was only isolated in 1983 (Barré-Sinoussi et al, 1983) which, according to the start date of 1959 invoked here, implies the virus circulated unnoticed for at least twenty-four years. During this interval, HIV managed to gain entry to most countries outside Asia and, thereby, exhausted many potentially long warning times. Moreover, the continued lack of a vaccine implies that those places, where these episodes are yet to lapse, will be unlikely to utilise the outstanding durations to mount immunisation campaigns.

4 Discussion

This paper has described a variety of mechanisms for representing either the mixing of specific cohorts of people at risk to HIV infection, or the passage of this virus between geographical regions. The former entails making choices about the degree of disaggregation of these populations that is necessary to capture the essential demographic influences on disease incidence; while the latter poses questions about the appropriate scale of analysis for detecting the diffusion of the agent. For most applications, however, these two sets of decision criteria are likely to be mutually exclusive. Detailed demographic categorisations, for example, have been best suited to analyses of communities occupying territories of limited geographical extent. In contrast, these subtle local variations might well be obscured by the specification of the extensive regions that tend to intersect with the slow progress of HIV in space and time (Lam et al, 1996). For these reasons, the highly disaggregated pandemic specification implied by model (20) is unlikely to be required in practice.

One of the most noticeable issues in recent debates about the applicability of HIV/AIDS modelling systems is the growing awareness of the importance of reproduction rates for understanding how transmission might be diminished. Health planning, for example, is assisted by knowing a particular community has an estimated rate of less than unity to imply a high degree of natural protection that might preclude the making of sustained interventions. Conversely, estimated rates in excess of unity point to the urgent need for action. In addition to this preventative role, these rates have made significant contributions to secular controversies about the idiosyncrasies of HIV incidence. The predominantly heterosexual mode of transmission which is most prevalent in central Africa (Earickson, 1990), for example, has often been hastily attributed to a high promiscuity rate among this population. Alternatively, high transmission probabilities of the order $\beta = 0.4$ estimated from samples of African patients (Peterman et al, 1988) have been linked the widespread prevalence of genital cuts and ulcers consequent upon prior infection with other STDs (Bassett and Mhloyi, 1991). This probability defines a very conservative critical partnership rate (equation 12) of $r^* = 1.25$ which, combined with the positive skew of most partner acquisition frequency distributions, is likely to place many more African heterosexuals at risk to infection than those elsewhere.

Such outcomes and arguments all imply that to account for national variations in the incidence of HIV/AIDS might require differently disaggregated multiregion models than those described here. In particular, region specific estimates of both β and λ are clearly needed to replicate the observed pathway of HIV and assess the differential impact of preventative actions. The implementation of such a specification would then pose new questions about how these mixing parameters interact to determine variations in the periodicity and spacing of HIV infection.

References

Anderson R M 1988 The epidemiology of HIV infection: variable incubation plus infectious periods in sexual activity *Journal of the Royal Statistical Society A* **151** 66-93

Anderson R M, May R M 1991 *Infectious diseases of humans: dynamics and control* Oxford: OUP

Bailey N J T 1975 *The mathematical theory of infectious diseases and its applications* High Wycombe: Charles Griffin

Bailey N J T 1995 Operational modelling of HIV/AIDS to assist public health control. In D Mollison (ed) *Epidemic models; their structure and relation to data* (Cambridge: CUP) pp.404-16

Bailey N J T, Estreicher J 1987 Epidemic prediction and public health control, with special reference to influenza and AIDS *Proceedings, World Congress of the Bernoulli Society* **1** 507-16

Banerjee K 1989 Rising prevalence of antibodies against HIV-1 in Western Maharashtra, India *Abstracts of the 5th International Conference on AIDS* Montreal Abstract TG022

Baroyan O V, Rvachev L A 1967 Deterministic epidemic models for a territory with a transport network *Kibernetika* **3** 67-74

Barré-Sinoussi F, Chermann J C, Rey F, Nugeyre N T, Charnaret S, Gruest J, Danduet C, Axler-Blin C, Vézinet-Brun F, Rouzioux C, Rozenbaum W, Montagnier L 1983 Isolating of a T-lymphotrophic retro-virus from a patient at risk from acquired immunodeficiency syndrome *Science* **220** 865-71

Bassett M T, Mhloyi M 1991 Women and AIDS in Zimbabwe: the making of an epidemic *International Journal of Health Services* **21** 143-56

Brookmeyer R 1991 Reconstructing the future trends of the AIDS epidemic in the United States *Science* **253** 37-42

Cliff A D, Haggett P, Ord J K 1986 *Spatial aspects of influenza epidemics* London: Pion

Cliff A D, Logan J D 1997 A structural time series approach to forecasting the space-time incidence of infectious diseases: post-war measles elimination in the United States and Iceland. In M M Fischer and A Getis (eds) *Recent developments in spatial statistics, behavioural modelling and neurocomputing* (New York: Springer) in press

Earickson R J 1990 International behavioural responses to a health hazard: AIDS *Social Science and Medicine* **31** 951-62

Flahault A, Valleron A 1992 A method for assessing the spread of HIV-1 infection based on air travel *Mathematical Population Studies* **3** 161-71

Gould P 1993 *The slow plague: a geography of the AIDS pandemic* Oxford: Blackwell

Gotlieb M S, Schroff R, Schanker H M, Weisman J D, Fan P T, Wolf R A, Saxon A S 1981 Pneumocystis carinii pneumonia and mucosal candidiasis in previously healthy homosexual men: evidence of a new acquired cellular immunodeficiency *New England Journal of Medicine* **305** 1425-31

Grant R M, Wiley J A, Winkelstein W 1987 Infectivity of the Human Immmunodeficiency Virus of homosexual men *Journal of Infectious Diseases* **156** 189-93

Haggett P 1994 Prediction and predictability in geographical systems *Transactions of the Institute of British Geographers* **19** 6-20

Haraldsdottir S, Gupta S, Anderson R 1992 Preliminary studies of sexual networks in a male homosexual community in Iceland *Journal of AIDS* **5** 374-81

Hethcote H W 1996 Modeling heterogeneous mixing in infectious disease dynamics. In V Isham and G Medley (eds) *Infectious diseases of humans: their structure and relation to data* (Cambridge: CUP) pp.215-38

Hethcote H W, Yorke J A 1984 Gonorrhoea transmission dynamics and control *Lecture Notes in Biomathematics* **56** New York: Springer

Heesterbeek J A P 1992 R_0 PhD thesis Leiden: University of Leiden

Isham V 1988 Mathematical modelling of the transmission dynamics of HIV infection and AIDS: a review *Journal of the Royal Statistical Society A* **151** 5-30

Jacquez J A, Simon C P, Koopman J 1991 The reproduction number in deterministic models of contagious diseases *Comments on Theoretical Biology* **2** 159-209

Jacquez J A, Simon C P, Koopman J 1995 Core groups and the R_0s for subgroups in heterogeneous SIS and SI models. In D Mollison (ed) *Epidemic models: their structure and relation to data* (Cambridge: CUP) pp.279-301

Kaplan E H 1989 Can bad models suggest good policies: sexual mixing and the AIDS epidemic *Journal of Sex Research* **26** 301-14

Kaplan E H, Lee Y S 1989 How bad can it get? Bounding worst case endemic heterogeneous mixing models of HIV/AIDS *Mathematical Biosciences* **99** 157-80

Kirp D L, Bayer R 1992 *AIDS in the industrialised democracies: passions, politics and policies* New Brunswick: Rutgers University Press

Knox E G 1986 A transmission model for AIDS *European Journal of Epidemiology* **2** 165-77

Knox E G, MacArthur C, Simons K J 1993 *Sexual behaviour and AIDS in Great Britain* London: HMSO

Lam N S-N, Fan M, Liu K 1996 Spatial-temporal spread of the AIDS epidemic, 1982-90: a correlogram analysis of four regions of the United States *Geographical Analysis* **28** 93-107

Li W H, Tanimura M, Sharp P M 1988 Rates and dates of divergence between AIDS virus nucleotide sequences *Molecular Biology and Evolution* **54** 313-30

Lin X, Hethcote H W, van den Driessche P 1993 An epidemiological model for HIV/AIDS with proportional recruitment *Mathematical Biosciences* **118** 181-95

Longini I M, Byers R H, Hessol N A, Tan W Y 1992 Estimating the stage-specific numbers of HIV infection using a Markov model and backcalculation *Statistics in Medicine* **11** 831-43

Longini I M, Fine P E M, Thacker S B 1986 Predicting the global spread of new infectious agents *American Journal of Epidemiology* **123** 383-91

Mariotto A 1989 Rate of growth of AIDS epidemic in Europe: a comparative analysis *Philosophical Transactions of the Royal Society, London B* **325** 175-8

May R M, Anderson R M, Blower S M 1989 The epidemiology and transmission dynamics of HIV/AIDS *Daedalus* **118** 163-201

Mollison D (ed) 1995 *Epidemic models: their structure and relation to data* Cambridge: CUP

Morris M 1991 A log-linear modeling framework for selective mixing *Mathematical Biosciences* **107** 349-77

Morris M 1995 Data driven network models for the spread of infectious disease. In D Mollison (ed) *Epidemic models: their structure and relation to data* (Cambridge: CUP) pp.302-22

Morris M 1996 Behaviour change and non-homogeneous mixing. In V Isham and G Medley (eds) *Infectious diseases of humans: their structure and relation to data* (Cambridge: CUP) pp.239-52

Peterman T A, Stoneburner R L, Allen J R, Jaffe H W, Curran J W 1988 Risk of HIV transmission from heterosexual adults with transfusion-associated infections *Journal of the American Medical Association* **259** 53-63

Raab G M, Gore S M, Goldberg D J, Donnelly C A 1994 Bayesian forecasting of the HIV epidemic in Scotland *Journal of the Royal Statistical Society A* **157** 17-30

Rosenberg P S, Gail H M, Carroll R J 1992 Estimating HIV prevalence and projecting AIDS incidence in the United States: a model that accounts for therapy and changes in the surveillance definition of AIDS *Statistics in Medicine* **11** 1633-55

Smyth F M, Thomas R W 1996a Preventative action and the diffusion of HIV/AIDS *Progress in Human Geography* **20** 1-22

Smyth F M, Thomas R W 1996b Controlling HIV/AIDS in Ireland: the implications for health policy of some epidemic forecasts *Environment and Planning A* **28** 99-118

Solomon P J 1996 AIDS: modelling and predicting. In V Isham and G Medley (eds) *Infectious diseases of humans: their structure and relation to data* (Cambridge: CUP) pp.407-38

Thomas R W 1988 Stochastic carrier models for the simulation of Hodgkin's disease in a system of regions *Environment and Planning A* **20** 1575-1601

Thomas R W (ed) 1990 *Spatial epidemiology* London: Pion

Thomas R W 1992 *Geomedical systems: intervention and control* London: Routledge

Thomas R W 1993 Source region effects in epidemic disease modeling: comparisons between influenza and HIV *Papers in Regional Science* **72** 257-82

Thomas R W 1994 Forecasting global HIV/AIDS dynamics: modelling strategies and preliminary simulations *Environment and Planning A* **26** 1147-66

Thomas R W 1996a Modelling space-time HIV/AIDS dynamics: applications to disease control *Social Science and Medicine* **43** 353-66

Thomas R W 1996b Alternative population dynamics in selected HIV/AIDS modeling systems: some cross-national comparisons *Geographical Analysis* **28** 108-25
Thomas R W 1997 Reproduction rates in multiregion modelling systems for HIV/AIDS *Journal of Regional Science* submitted for publication
Thomas R W, Smyth F M 1997 Estimating the impact of preventative action on the space-time incidence of HIV/AIDS. In M M Fischer and A Getis (eds) *Recent developments in spatial statistics, behavioural modelling and computational intelligence* (New York: Springer) in press
Van Druten J A M, Reintjes A G M, Jager J C, Heisterkamp S M, Poos M J J C, Countinho R A, Dijkgraaf M G W, Ruitenberg E J 1990 Infection dynamics and intervention experiments in limited risk groups *Statistics in Medicine* **9** 721-36
Wilson A G, Bennett R J 1985 *Mathematical models in human geography and planning* Chichester: Wiley
Yorke J A, Hethcote H W, Nold A 1978 Dynamics and control of the transmission of gonorrhoea *Sexually Transmitted Diseases* **5** 51-156

A New Approach to Space-Time Modelling of Infectious Disease Behaviour

Andrew B. Lawson*& Petra Leimich†
Mathematical Sciences Division,
University of Abertay Dundee,
Bell Street, Dundee,UK

September 26, 1997

Abstract

A new approach to the space-time modelling of infectious diseases is considered. A modulated heterogeneous Poisson process with intensity defined as a function of a 2-dimensional susceptibility field is proposed. The model is fitted to a measles epidemic.

1 Introduction

Our aim in this paper is to address both the incorporation of heterogeneous population background and the modelling of spatio-temporal spread of infectious disease. To do this we borrow some ideas from the recent developments in the modelling of non-infectious diseases. In particular, in studies of non-infectious disease it is commonplace to incorporate population heterogeneity within models for disease distribution(see, e.g. [9],[8]). We term the heterogeneous population as the 'at risk' background. This corresponds with the usual definition of the susceptible population i.e. the population of susceptibles are those who are 'at risk' at any specified point in space-time of getting infected by the disease of concern. In addition, the formulation of models for the space-time behaviour of non-infectious disease relies on the specification of components which depend directly on, or are modified by, this function of susceptibles. This parallels the development of models for infectious disease (see

*email: a.lawson@tay.ac.uk
†email : p.leimich@tay.ac.uk

e.g. [2], Ch.6). In addition, the connection between the modelling of *clusters* of disease where the aetiology is unknown or uncertain (e.g. leukaemias) leads to the consideration of infectious agents and hence overlaps with infectious disease modelling. While modelling of disease clusters *per se* can be achieved without recourse to models for infectious behaviour, it is reasonable to assume that spatial and temporal clustering can be modelled explicitly via a form of contact probability field which will lead to clustering in space-time. This field can be derived from purely descriptive models for spatial clusters of disease (see e.g. [5],[6],[7]).

2 Model Development

In what follows we assume that a realisation of n disease events occurs within a fixed spatial and temporal window. We denote these windows as U and T respectively. The disease events are cases of infection and hence, $\{\mathbf{x}_i, t_i\}, i = 1, ..., n$ represents the locations and infection times of all the cases. Now at any specified time t_*, there will be a finite number of infectives, who have the potential to convert susceptibles to infected cases. Denote the set of infectives at t_* as $I(t_*) : \{\mathbf{x}_{I_j}, t_{I_j}\}$ $j = 1, ..., n_{t_*}$. We assume that the probability of any susceptible being infected is related to the set $I(.)$, and hence we construct our model around dependence on the current infective set at any time. In previous work on such models in the temporal domain, the basic assumption is made that the incidence of infection is a simple product of susceptible number and infective number. However, to make the dependence specific for spatial and temporal locations, it is convenient to specify a more detailed model of this association. First, we specify the form of the susceptible population. As this population will be spatially and temporally variable, we introduce a three dimensional field representation $S(\mathbf{x}, t)$ which represents the degree of local susceptibility in the population at $(\mathbf{x}, t)$. This specification of the susceptible population can be seen as a general method which can make allowance for discrete susceptible locations (e.g. houses) or more continuous backgrounds (e.g. urban areas). In the case of discrete locations, $S(\mathbf{x}, t)$ will have a series of spikes at those locations. This definition of the susceptibility function mirrors the use of such a function for non-infectious diseases. In that case, $S(\mathbf{x}, t)$ is often estimated from standardised rates for the community, given the local population (age-sex) structure (see e.g. [4]). In studies of infectious spread where the infection arises within a large population, $S(\mathbf{x}, t)$ could be estimated nonparametrically via density estimation (see e.g.[12]).

First, we assume that we can model the disease process at any time, given knowledge of the current state of the infective population. To do this, we assume that the first order intensity of cases can capture the model structure adequately, and hence the incidence of cases, conditional on the current $I(.)$, can be modelled via a modulated heterogeneous Poisson process with first order intensity:

$$\lambda(\mathbf{x},t) = \rho.S(\mathbf{x},t).\sum_{j=1}^{n_t} h(\mathbf{x}-\mathbf{x}_{I_j}).g(t-t_{I_j}) \tag{1}$$

where $S(\mathbf{x},t)$ is the local density of susceptibles at $(\mathbf{x},t)$, ρ is the overall density (space×time units), h is a spatial cluster function which relates the location of a susceptible to any current infective location, n_t is the current number of infectives (at time t). The g function is a cluster function depending on the temporal position (t) in relation to the time of infectivity of the known infectives (t_{I_j}). This can be structured to model special temporal infectivity periods (e.g. prodromal duration in measles). The h, g functions will usually have a distance decay form i.e. they may produce lower intesity the further away from the location and time of the infective events the location is. The temporal function can include an infectivity period and other forms. This specification of the first order intensity relates the local density of susceptibles to their spatial and temporal distance from currently infective people.

3 Modelling Special Cases

In a later section we will consider some modifications to this model, in an application to a German measles epidemic, reported by [11] and recently revisited by [10],[1] and [3]. However, before we discuss these specific modifications, it is worthwhile considering a special case of this model and the resulting simplifications.

3.1 Proportional Hazards Interpretation

Given the temporal nature of this problem, in which events occur at observed time points, it is interesting to pursue the connection between this modelling approach and conventional survival analysis. In the proportional hazards model, a risk set is observed over time and any failures (disease cases) are assessed conditional on the risk set R specified just prior to the failure time of the individual of concern. A similar development can be pursued here. If we regard (1) as the hazard function for a disease case, then we can specify the probability of an infection within $R(t)$ within a small time increment δt

approximately as

$$\sum_{R(t)} \lambda(\mathbf{x}, t)\delta\mathbf{x}\delta t$$

where $\delta\mathbf{x}$ is a small area around $\mathbf{x}$. Hence we can also specify a conditional probability of a particular individual at $\mathbf{x}_i$ becoming infected as:

$$\frac{\lambda(\mathbf{x}_i, t)\delta\mathbf{x}\delta t}{\sum_{R(t)} \lambda(\mathbf{x}, t)\delta\mathbf{x}\delta t}.$$

If it is assumed that the δ terms cancel, we can take the product of these conditional probabilities evaluated at the case infection times to give a conditional likelihood:

$$L = \prod_i \frac{\lambda(\mathbf{x}_i, t_i)}{\sum_{j\in R(t_i-)} \lambda(\mathbf{x}_j, t_i)} \tag{2}$$

where $R(t_i-)$ denotes the risk set just prior to infection time t_i. In general, the background susceptible function will *not* factor out of this likelihood and so there is still a requirement to estimate the susceptible function directly. Note that direct maximisation of (2) is possible and this could avoid the evaluation of integrals over spatial and temporal domains required by the Poisson process likelihood formulation.

4 Cluster Function Specification

In what follows we consider in more detail the definition of the cluster functions h and g. These functions determine the contact relationships between potential cases and the existing infected population.

4.1 Spatial dependence

The spatial dependence function $h(u)$ can take a variety of forms depending on the choice of contact distribution specified. The simplest forms are those which assume that u is a simple distance measure relating a case residence $(\mathbf{x})$ to the residences of infectives $(\mathbf{x}_{I_j})$. In that case the definition of h can reduce to a function of distance between residences. Here the inter-residence distance is assumed to form a surrogate for exposure. This may be reasonable for certain types of disease, where contact occurs via 'local' behaviour. Where special contact patterns are important (e.g. with Aids), inter-residence distance may not be a useful surrogate.

A typical spatial dependence cluster function is defined in Section 5.2.

4.2 Temporal dependence

The spatial interaction discussed above is directly modified by the temporal cluster function in (1). This modification implies that even when strong spatial association is present, if weak temporal association is present, there will be a reduced probability of infection. This appears to be a realistic assumption for most infectious diseases. Often it is useful to consider a model for the temporal infection process in an individual and to base a $g(.)$ function on this specification. A typical profile of infection can be broken into three stages: a period of incubation, an infectious period, and a final period. Often the final period is represented by *removal* of the susceptible from the population, if the disease is such that after infection there is little or no probability of contracting the disease again. This type of model is often referred to as a SIR model: susceptible-infected-removed model.

A typical specification for $g(.)$ is then:

$$g(t - t_I) = \begin{cases} f_1(t) & \text{if } t < t_{I_0} \\ f_2(t) & \text{if } t_{I_0} \leqslant t < t_{I_1} \\ f_3(t) & \text{if } t \geqslant t_{I_1} \end{cases}$$

where the f_i functions apply to different periods and t_{I_0} and t_{I_1} are the start and end times of the infectious period.

5 Data Example

We will now consider the spread of a measles epidemic, described by [11] and [10]. This epidemic occurred within a small isolated village, Hagelloch, Germany, in 1861, effectively a closed community. The data set is unusually complete, as [11] meticulously recorded the progress of the epidemic. On a daily basis the household and name of the family members affected was recorded, including the start, development and disappearance of the various symptoms, body temperature, and any complications or deaths. There is a complete record of all susceptibles. [10] mapped the locations of susceptibles and cases in space, and established the most likely infector for each susceptible that became infected.

The population of the village at the time of the epidemic comprised 577 inhabitants. There were 200 children up to the age of 15, who were born after the previous measles epidemic or escaped infection as infants. Twelve of these can be regarded as not susceptible as they were immigrants who had had the measles before, infants aged 6 months or less (carrying placental immunity), or were kept in isolation. The remaining 188 susceptible children were infected.

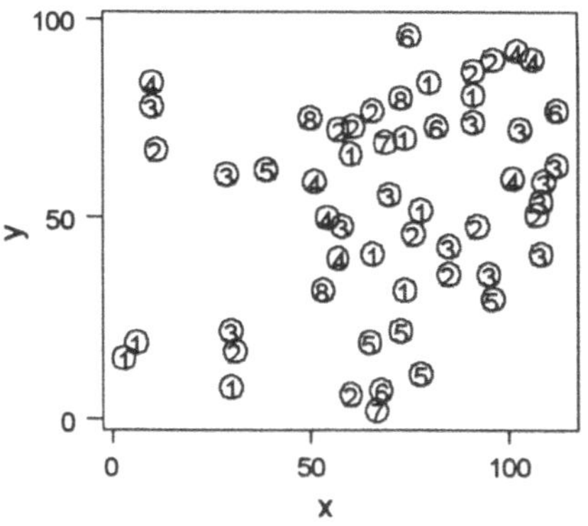

Figure 1: Map of Hagelloch, showing the number of susceptibles in each household before the start of the epidemic.

To apply the proportional hazards model (1,2) to this data set, we first specify the functions contributing to the first order intensity (1).

5.1 Distribution of Susceptibles $S(\mathbf{x}, t)$

The number of susceptibles at location $\mathbf{x}$ at time t, $S(\mathbf{x},t)$ is known from the data (see Fig. 1). Time in the model is discrete, as observations were made daily. The grid of locations given by [10], which was approximately 100×100 units, was scaled to 1×1. To obtain a continuous surface, the data at each time t was replaced by a bivariate Gaussian smoother $\hat{S}(\mathbf{x},t)$. The smoothing parameter was calculated using Silverman's [12] rule of thumb formula $0.96\sqrt{\frac{1}{2}\left(\text{var}(x_i) + \text{var}(y_i)\right)n^{-1/6}}$. The smoothed surface obtained for the susceptibles before the start of the epidemic ($t = 0$) is shown in Fig. 2.

The spatial and temporal distance functions g(.) and h(.), describe the contact relationships between the susceptibles and the existing infective population.

5.2 The Spatial Distance Function h

For diseases easily transmitted through general contact, the spatial distance between residences can be used as a measure of exposure, and a bivariate normal function used:

$$h\left(\mathbf{x} - \mathbf{x}_I\right) = \frac{1}{2\pi\kappa} \exp\left(-\frac{1}{2\kappa}\left(\|\mathbf{x} - \mathbf{x}_I\|\right)^2\right) \tag{3}$$

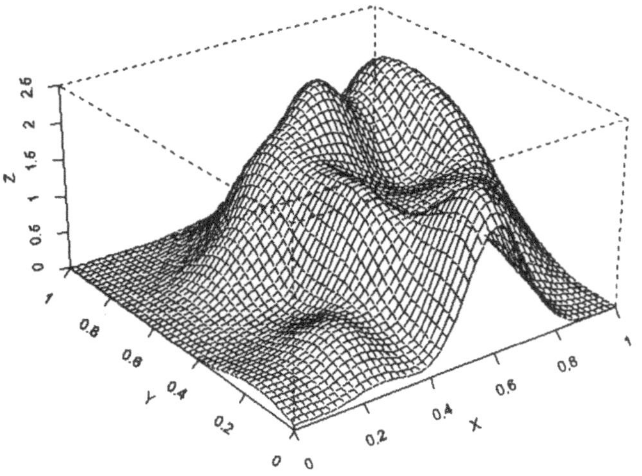

Figure 2: Smoothed density of susceptibles $\hat{S}(\mathbf{x}, 0)$.

stage	suspect.	incubation		prodrome	eruption	desquamation
duration		variable	1 day	≈4 days very variable	≈3 days variable	
		14 days, almost constant				
status	suspect.	latent	infectious			removed

Table 1: Summary of the measles infectivity pattern.

where $\mathbf{x}$ denotes the location of a susceptible, $\mathbf{x}_I$ the location of an infective. The one parameter κ is a spread parameter; the larger κ, the more likely is infection across some distance.

5.3 The Function g

The function $g(.)$ describes the changes of infectivity over time. Its specification is based on the infectivity pattern of measles, summarised in Table 1.

The times of the start of the prodrome (first symptoms), of the eruption (rash), and, if applicable, of death, are available for each infective in the data set as PRO, ERU and $DEAD$. Following [10, 11], we assume individuals to be equally infectious from a day before the start of the prodrome until three days after the eruption and define

$$t_I = PRO - 1,\ t_R = \min\{ERU + 3, DEAD\}.$$

For $g(.)$, we can then use a simple uniform function

$$g(t; t_I, t_R) = \begin{cases} \gamma & \text{if } t_I < t < t_R \\ 0 & \text{otherwise} \end{cases} \tag{4}$$

rather than the more general trapezoidal function shown in Fig. 3. The parameter γ is a measure of the infectivity.

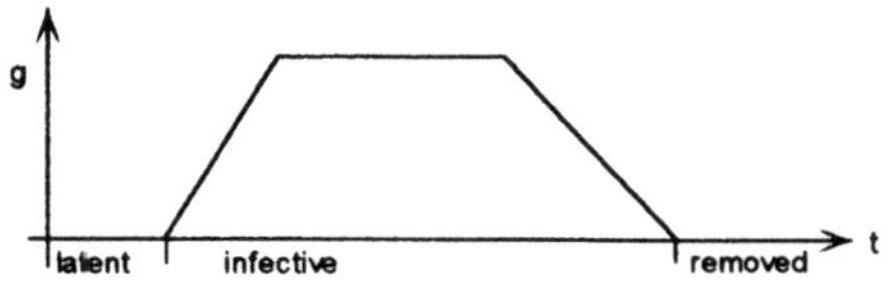

Figure 3: A typical function $g(.)$

5.4 Fitting the model

Substituting into (1) and (2), we obtain

$$L = \prod_i \frac{\widehat{S}(\mathbf{x}_i, t_i) \sum_{j=1}^{n_{t_i}} g(t_i; t_{I_j}, t_{R_j}) h(\mathbf{x}_i - \mathbf{x}_{I_j})}{\sum_{j \in R(t_i -)} \widehat{S}(\mathbf{x}_j, t_i) \sum_{k=1}^{n_{t_i}} g(t_i; t_{I_k}, t_{R_k}) h(\mathbf{x}_j - \mathbf{x}_{I_k})}. \tag{5}$$

Noting that the value of $g()$ is γ wherever it is used in (5), further simplification of (5) yields

$$L = \prod_i \frac{\widehat{S}(\mathbf{x}_i, t_i) \sum_{j=1}^{n_{t_i}} h(\mathbf{x}_i - \mathbf{x}_{I_j})}{\sum_{j \in R(t_i -)} \widehat{S}(\mathbf{x}_j, t_i) \sum_{k=1}^{n_{t_i}} h(\mathbf{x}_j - \mathbf{x}_{I_k})}. \tag{6}$$

To fit the model to the data set, the log-likelihood was maximised with respect to the single remaining parameter κ. This gave very large values for κ, see Fig. 4(a). The disappointing results can be explained readily by returning to the particular features of the data example. Many of the susceptibles attend the village school. The spatial distance function $h(.)$ currently ignores this likely place of infection, as it is based only on the distance between residences. Therefore, we would expect the model to improve, as observed, with a flattening of the distance function h, which occurs as κ increases.

6 Revised Model

The school has two classes, one for 6-10 year-olds and one for older children. The school status is known for all susceptibles. A good model should incorporate the school in the sense that infection is likely to take place at school between classmates, and to a lesser extent between children in different classes.

To implement this, the spatial distance function h is modified. There are several possible approaches, such as making κ a function of the age group, or using a non-euclidean measure of distance. We adopt the second approach, defining the modified function h as

$$h\left(\mathbf{x}-\mathbf{x}_I\right)=\frac{1}{2\pi\kappa}\exp\left(-\frac{1}{2\kappa}\left(\|\mathbf{x}-\mathbf{x}_I\|\times m\right)^2\right) \tag{7}$$

where m modifies the distance according to school status:

$$m=\begin{cases}1 & \text{if susceptible and infective not both at school}\\ 0 & \text{if susceptible and infective in same class}\\ \frac{1}{2} & \text{if susceptible and infective in different classes}\end{cases}$$

6.1 Results

Maximising the log-likelihood for the revised model, using the Splus function optimize(), yielded $\kappa_{\text{opt}}=0.0337$ after convergence, see Fig. 4(b). Using the general result $2\times\left(L_{\kappa_{\text{opt}}}-L_\kappa\right)\sim\chi_1^2$ to find a 95% likelihood confidence interval for κ, we obtain CI $=(0.0205, 0.0740)$.

Finally, we used an empirical approximation based on sampling the likelihood surface to provide a variance estimate for κ. Using rejection sampling from the surface to provide a sample of a hundred λ values, we computed the empirical standard deviation, $\widehat{sd}(\kappa)=0.020321$.

7 Conclusions

We have presented here a general approach to the modelling of infectious disease behaviour which can easily be applied to a range of data formats, whether

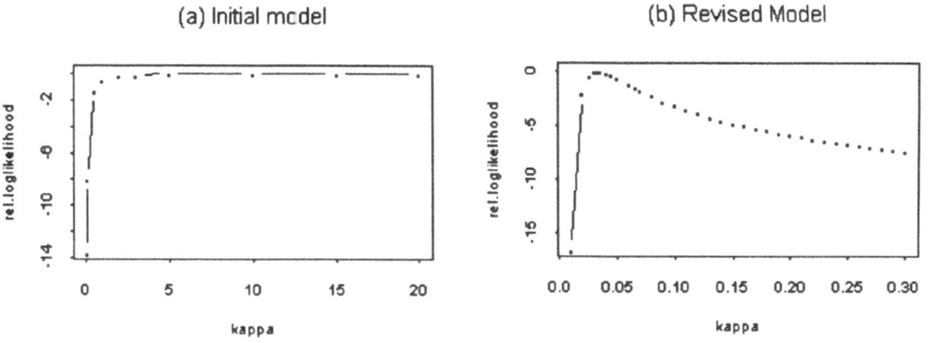

Figure 4: Plots of the relative log-likelihoods for a range of κ values. (a) Original model; (b) revised model with school.

in the form of case addresses of infectives and susceptibles, or in the form of counts of infectives within regions in fixed time periods with lower level susceptible information. We have demonstrated from our example that the modelling approach is easily applied with standard statistical packages and does not require extensive programming.

Our example has been analysed using a proportional hazards type of model and we have been reasonably successful in demonstrating the importance of school contacts in the spatial contact process. We need to carry out further work in the application area and other examples to further refine our model and its components.

Finally, it is worth noting that our modelling approach can be extended to include situations where the infective locations are unknown and need to be estimated from the observed data. This area can be examined by the use of *data augmentation* methods using computational tools such as McMC (see e.g. [13]).

References

[1] P. Aaby, H. Oesterle, and K. Dietz. Severe outbreak of measles in an isolated German village, 1861: I: Mortality and secondary attack rates, epidemiology and infection. *submitted*, 1993.

[2] N. G. Becker. *Analysis of Infectious Disease Data*. Chapman and Hall, 1989.

[3] N. G. Becker and D. G. Wang. Severe outbreak of measles in an isolated German village: II: Analysis of transmission rates. *submitted*, 1993.

[4] H. Inskip, V. Beral, P. Fraser, and P. Haskey. Methods for age-adjustment of rates. *Statistics in Medicine*, 2:483–493, 1983.

[5] A. B. Lawson. Markov chain Monte Carlo methods for putative pollution source problems in environmental epidemiology. *Statistics in Medicine*, 14:2473–2486, 1995.

[6] A. B. Lawson. Markov chain Monte Carlo methods for clustering in case event and count data in spatial epidemiology. *ISCB invited paper: submitted*, 1997.

[7] A. B. Lawson. Some spatial statistical tools for pattern recognition. In A. Stein, F. W. T. P. de Vries, and J. Schut, editors, *Quantitative Approaches in Systems Analysis*, volume 7, pages 43–58. C. T. de Wit Graduate School for Production Ecology, 1997.

[8] A. B. Lawson and N. Cressie. Spatial statistical methods for environmental epidemiology. In C. R. Rao and P. K. Sen, editors, *Handbook of Statistics:Bio-Environmental and Public Health Statistics*. 1998. to appear.

[9] A. B. Lawson and L. Waller. A review of point pattern methods for spatial modelling of events around sources of pollution. *Environmetrics*, 7:471–488, 1996.

[10] H. Oesterle. *Statistische Reanalyse einer Masernepidemie 1861 in Hagelloch.* PhD thesis, Eberhard-Karls-Universität Tübingen, 1990.

[11] A. Pfeilsticker. *Beiträge zur Pathologie der Masern mit besonderer Berücksichtgung der Statistischen Verhältnisse.* PhD thesis, Eberhard-Karls-Universität Tübingen, 1863.

[12] B. W. Silverman. *Density Estimation for Statistics and Data Analysis.* Chapman and Hall, London, 1986.

[13] M. A. Tanner. *Tools for Statistical Inference.* Springer verlag, New York, 3rd edition, 1996.

Spatial modelling of cluster object and non-specific random effects, with application in spatial epidemiology

Andrew B Lawson
Mathematical Sciences Division
University of Abertay Dundee
Bell Street
Dundee
U.K.

Abstract

The spatial modelling of small area health data has, for some time, included spatial autocorrelation as a random effect. This effect is non-specific and global and does not address the location of clusters of disease (a specific task). This paper addresses the need for specific and non-specific random effects within spatial epidemiology. In addition, individual frailty is also considered important and computational algorithms based on a range of MCMC methods are described.

1 Introduction

The modelling of the spatial distribution of events often involves the consideration of random effects which can account for unobserved heterogeneity in such data. This heterogeneity can be uncorrelated (such as pure overdispersion in count data) and/or spatially correlated. In the case of small area health data, a considerable literature has grown around the modelling of such effects for counts of disease found in arbitrary regions (e.g. census tracts)(see e.g. [2][3]). To a lesser extent, these random effects have been modelled in case event data (see e.g. [13] and for a general review).

Most models for random effects in such data regard the effect to be modelled as non-specific. That is, no specific spatial locational structure is to be estimated, and the effects only relate to a global measure of heterogeneity. An alternative approach to the modelling of random effects is to include specific locational structure within the random effects to be considered. This locational structure could relate to unobserved spatial features, and the aim of the study is to recover or estimate this unobserved set of spatial features. The inclusion of such random *object* effects in a modelling approach compliments the non-specific random effect modelling and also unifies the idea of cluster modelling and conventional random effect modelling, mirroring the duality in image processing between low level image restoration and higher level object recognition.

In this paper, an approach to modelling both types of random effect is described in an application to modelling the spatial distribution of childhood Leukaemia and Lymphoma cases in Humberside. This data set was first published by Cuzick and Edwards[7]. Figure 1 and 2 display the case addresses of the disease of interest and also the map of a 'control' realisation. In this case a random sample of addresses of births within the study window was taken. Cuzick and Edwards discuss the use of such controls in the representation of local population 'at risk'. In this case the local birth rate is intended to represent variation in population density, which must be incorporated into a model for disease within a heterogeneous and spatially-varying environment.

2 Model Development

We assume that cases of disease are regarded as arising independently from within the local population. It is assumed that the cases follow a modulated heterogeneous Poisson process with first order intensity given by:

$$\lambda(\mathbf{x}) = g(\mathbf{x}) . f(\mathbf{x}; \boldsymbol{\theta}), \tag{1}$$

where $\mathbf{x}$ denotes the spatial location , $g(.)$ is a function of the 'at risk' population background, and $f(.)$ is a function of location and a vector of parameters $\{\theta\}$. Here f will be parameterised with a variety of model elements which include terms relating to clustering of the disease, spatial trend and also background individual random effects (frailties). These individual effects are termed *frailties* as they represent individuals random variability of susceptibility to disease. Note that a likelihood for the observed case event data can

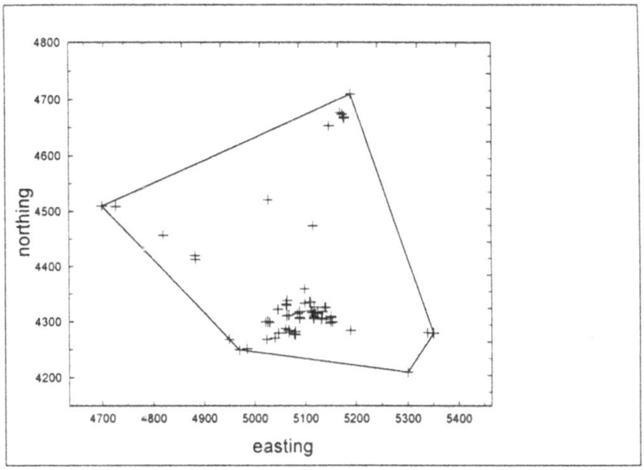

Figure 1: Case event map

be immediately defined:

$$L_1 = \frac{1}{m!} \prod_{i=1}^{m} \lambda(\mathbf{x}_i) \,.\, \exp\left\{ - \int_T \lambda(\mathbf{u})\, d\mathbf{u} \right\} \tag{2}$$

where $\mathbf{x}_i, i = 1, m$ are a realisation of case events within a fixed window T. In 2 with the intensity 1, it is important to be able to estimate $g(.)$, or to reformulate the model and condition out $g(.)$ as it is a nuisance function. A general method of accounting for $g(.)$ and its variability, is to regard the smoothing constant in a nonparametric estimate of $g(.)$ as a random variable which has a prior distribution, and during estimation of components in f the smoothing constant is allowed to vary. We adopt this approach here, and include a Metropolis -Hastings (MH) step within our overall algorithm which generates new smoothing values.

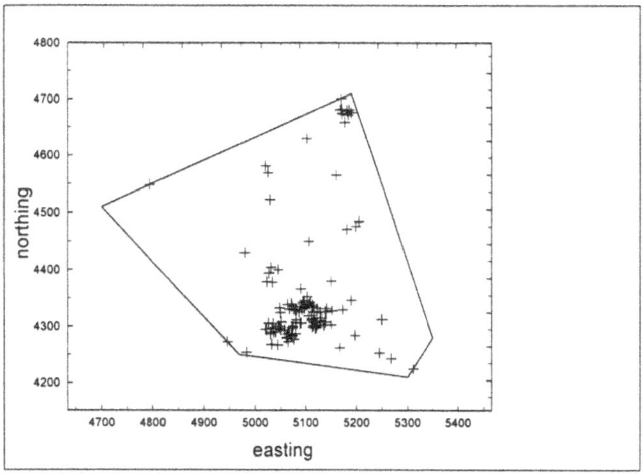

Figure 2: Control event map

3 The $f(.)$ structure and Prior Distributions

Our main focus in this work is the modelling of cluster structure within epidemiological data. This task is approached here by the structuring of f to describe the possible sources of variation in the data.

It is assumed that a set of n clusters exist within the data and these can be described by a cluster distribution function which relates the case events to unobserved cluster centres $\{\mathbf{x}_c\}_{j=1,n}$. Here we assume that the distribution function has the form:

$$1 + \sum_{j=1}^{n} h(\mathbf{x} - \mathbf{x}_{c_j}) \tag{3}$$

where some of the n centres could lie outside T, within a larger area U. The use of such an external boundary allows external centres to be estimated and hence edge effects can be largely circumvented for cluster terms. Lawson[11], and Baddeley and van Lieshout[1] and subsequently [12] have discussed the derivation and use of this approach to clustering via the use of prior distributions for $\{\mathbf{x}_c\}$ and other parameters, with associated posterior sampling based on MCMC algorithms. The effects to be included are regarded as nuisance effects related to possibly unobserved heterogeneity in the data , but not accounted for by the $g(.)$ background function. Clayton[5] has discussed the use of frailty models , where individual variation in susceptibility to disease is modelled by a random effect. This frailty concept can be applied in the case event context also and is included in our formulation.

A flexible approach to modelling such structure is to parameterise these effects for the i th individual as:

$$\zeta_i = \exp\{t_i + u_i + v_i\} \tag{4}$$

where t_i is a spatial trend term, u_i is a spatially correlated heterogeneity term and v_i is an uncorrelated heterogeneity term, following [2]. In the example that follows we do not include a trend term,but this could be used in other examples, and of course, it could include regression terms which are functions of spatial variables or covariates. In our case, we have included prior distributions for the u and v terms which extends the ideas of Besag and coworkers[2]. The Intrinsic Autoregressions improper difference prior distribution developed from the lattice models of Kunsch[10], can be extended to allow a distance parameterisation for the covariance structure. Hence the prior for $\{u\}$ is defined as:

$$p_i(u_i|.....) \propto \exp\{-\sum_{k\in\partial_i} w_{ij}(u_i - u_j)^2\} \tag{5}$$

where w_{ij} is a distance weight function , and ∂_i is a defined neighbourhood of the i th point. In the following , we assume $w_{ij} = \exp(-d_{ij})/\{2r\}$, where r is a range parameter . The neighbourhood ∂_i is assumed to be a fixed distance and in the example used here it is taken as half the maximum dimension of the window T. Note that the weights used here mimic the weights used in fully specified covariance priors. Edge correction has not been employed to this weighting system, and hence some distortion will occur close to the boundaries of T, although for this example this is unlikely to be severe. The uncorrelated heterogeneity (v_i) is defined to have a conventional Gaussian prior as in [2]:

$$p(\mathbf{v}) \propto \sigma^{-m/2} \exp\left\{-\frac{1}{2\sigma}\sum_{i=1}^{m} v_i^2\right\}. \tag{6}$$

Both r and σ are assumed to have improper inverse exponential hyperpriors:

$$prior(r,\sigma) \propto e^{-\epsilon/2r} e^{-\epsilon/2\sigma}, \qquad \sigma, r \succ 0, \tag{7}$$

where ϵ is taken as 0.001. These prior distributions penalise the absorbing state at zero, but provide considerable indifference over a large range.

With these ingredients the joint posterior distribution can be specified:

$$P_o(\boldsymbol{\theta}|\mathbf{x}) \propto \mathbf{L}_1.p(\mathbf{v}).p(\mathbf{u}).\text{prior}(r, \sigma).p_h \quad (8)$$

where p_h represents priors for terms in $h(.)$ and

$$\lambda(\mathbf{x}_i) = g(\mathbf{x}_i).\zeta_i.\{1 + \sum_{j=1}^{n} h(\mathbf{x}_i - \mathbf{x}_{c_j})\}.$$

The marginal posterior distribution of cluster centres is the main focus and to sample this effectively it is also important to specify a structure for $h(.)$ and associated prior distributions. The cluster centres and the number of centres are unknown and so it is appropriate to assume a prior distribution for both centre locations and number of centres. Following previous work, we assume a Markov inhibition prior (Strauss prior) for the centres (see e.g.[12]). This prior contains two parameters, α and $\mathbf{R}$, governing the inhibition process, where the probability of a realisation is

$$p(\mathbf{x}_c) \propto \rho\alpha^{n_{\mathbf{R}}},$$

where ρ is an overall rate parameter, $n_{\mathbf{R}}$ is the number of $\mathbf{R}$ *−close* pairs of points in the configuration, and $\mathbf{R}$ is a specified inhibition distance. This controls the density of centres by prohibiting centres within a certain distance of other centres. We also assume a Gaussian radial cluster distribution:

$$h(\mathbf{x} - \mathbf{x}_c) = \frac{1}{2\pi\kappa} e^{-\{||\mathbf{x}-\mathbf{x}_c||\}^2/2\kappa}$$

where κ is a cluster parameter. The hyperpriors for ρ, κ and α are uniform on suitable ranges. The parameter $\mathbf{R}$ is closely correlated with the inhibition parameter α, is fixed at a suitable (large) range to avoid indeterminacy. Figure 3 displays the directed acyclic graph (DAG) for the overall model specification.

4 Posterior Sampling via MCMC

The full posterior distribution specified above can be sampled using conventional MCMC methods for most parameters. The heterogeneity hyper

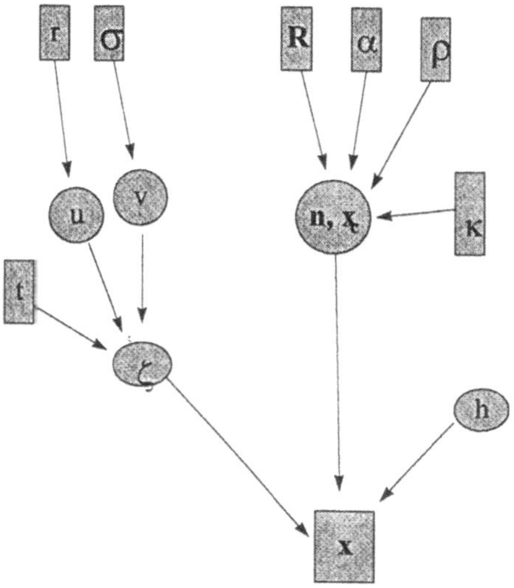

Figure 3: Directed Acyclic Graph of cluster model

parameters (r, σ) are still within the inverse Gamma family and hence Gibbs steps can be used to update these parameters. Other parameters can be sampled via a variety of MH and Gibbs steps in a conventional way, except for the cluster centre parameters $\{\mathbf{x}_c, n\}$. These parameters form part of a mixture where the upper limit of the mixture sum is unknown. While it is possible to approximate the associated conditional distributions and apply a Gibbs Sampler (see e.g. [11]), this problem is well suited to joint sampling of the $\{\mathbf{x}_c, n\}$ distribution via reversible jump sampling ([9]). This sampling method is described in detail in [12]. The method consists of iterative evaluation of posterior ratios for the addition/deletion and shifting of cluster centres and produces a Metropolis-Hastings algorithm termed the SBDS (spatial-birth-death-shifting) algorithm.

In the following example, we run the SBDS algorithm, conditional on all other parameters, with $U = T$ (to allow for seaward boundaries), and equal probability birth and death transitions with no shifting. The death rate is assumed to be $1/n$ and uniform birth rate in a disc of radius 0.01. The sampler is run for a single iteration at each step of the main sampler algorithm. The main sampler algorithm was run to convergence within 20000 iterations. Convergence was assessed by a variety of methods, including variance summaries and graphical methods (qq plots), as suggested by a variety of authors. A review of convergence diagnostics is provided in [6].

Model selection criteria must be used to assess which, if any, of the chosen models are to be preferred. In our examination of model differences our primary focus is on the comparison of conventional heterogeneity parameters and the inclusion of random object (cluster) model components. To this end we examine 3 different model formulations: 1) a full model with both ζ_i and cluster distribution function (3), 2) a partial model with ζ_i only, and 3) a partial model with the cluster distribution function (3) only. These alternative models consist of 7, 3 and 5 parameters respectively. Comparison of different models within a Bayesian modelling approach can be achieved by the use of posterior predictive inference. In particular, the Bayesian Information Criterion(BIC) can be used to assess whether samples from the posterior distribution support one or other of the models examined. This measure is defined as:

$$2\widehat{l(\theta^*)} - \frac{1}{G}\sum_{k=1}^{G} p_k \log n$$

where θ^* is defined to be a posterior sample of parameters, p_k is the k th number of parameters, and

$$\widehat{l(\theta^*)} = \frac{1}{G}\sum_{k=1}^{G} l(\theta^*_{(k)})$$

the average likelihood evaluated over G posterior samples (see [4] for discussion). BIC values for different models can be compared and a simple comparison of their values made, the highest value being favoured.

5 Data Example

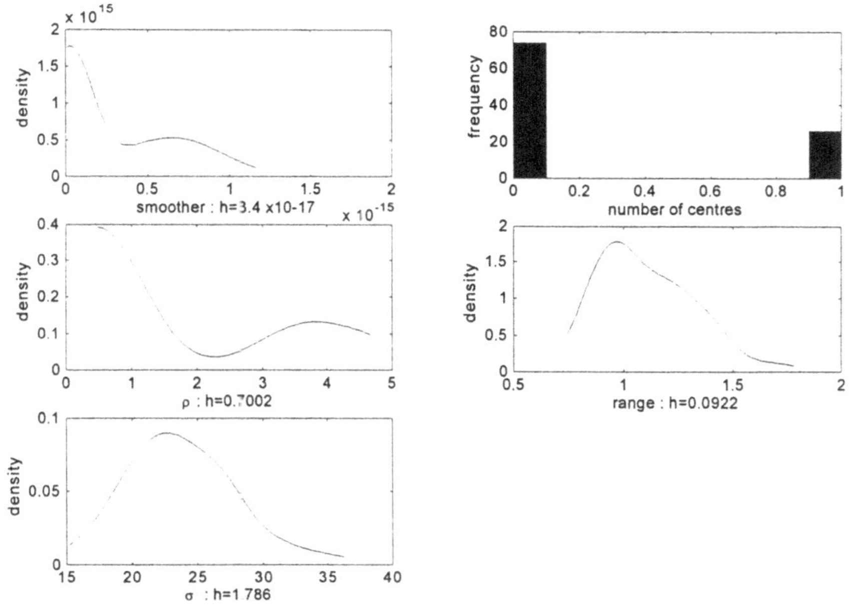

Figure 4: Posterior marginal distributions of parameters

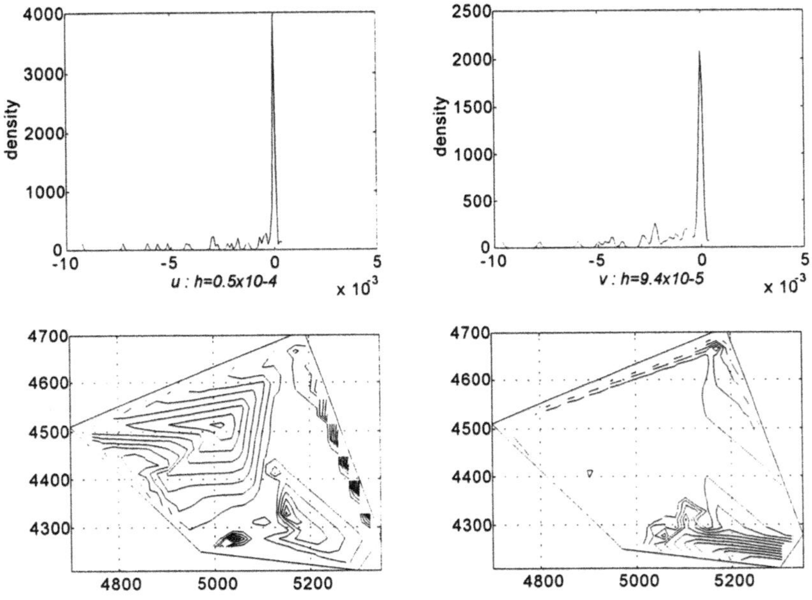

Figure 5: Posterior marginal distributions of u and v.

Figure 4 displays the posterior marginal distributions for parameters in model 1) : the full model. Figure 5 displays the posterior marginal distributions for the correlated and uncorrelated heterogeneity for model 2. The BIC values for the three models are :

model	BIC
1) full	-320.32
2) ζ_i only	-284.30
3) (3) only	-311.73

The primary conclusion for this example is that, based on the BI criteria, model 2 is favoured. However, all models show relatively low predictive probability. The full model displays (figure 4) suggest that the mode for number of clusters is zero, and the posterior marginal of cluster centres is relatively uniform, and this supports the redundancy of the cluster sum term (3), as indicated from the BIC analysis. The marginal distribution of ρ (rho) confirms this result.. There is therefore strong evidence for no clustering in this data set, beyond that described by heterogeneity. The heterogeneity found is of limited extent but with marked differences in spatial structure between uncorrelated and correlated forms (figure 5). The scaling effects of the r (*range*) and σ (*sigma*) are apparent. Overall, this analysis supports the previous findings of [7], and [8] who noted little evidence for clustering in this data set.

6 Conclusions and Future work

Further work is required in the analysis of sensitivity of results to the choice of prior distributions and in the choice of components of heterogeneity. In addition, the examination of predictive residuals would help to assess pointwise goodness-of-fit. Note that the inclusion of edge compensation/correction for the weighting system (w_{ij}) would also be advantageous.

Overall this paper demonstrates the possibility of incorporating random effects of varying kinds in the analysis of spatial clustering in small area epidemiology. These effects can capture spatial correlation, conventional uncorrelated frailty and spatial trend and/or individual covariates. Note also that it is straightforward to extend this analysis to regionalised count data where region centroids are known, or where digitised boundaries are available.

References

[1] A. Baddeley and M. Lieshout. Stochastic geometry models in high-level vision. In K. Mardia, editor, *Statistics and Images*, pages 233–258. Carfax, Abingdon, 1993.

[2] J. Besag, J. York, and A. Mollié. Bayesian image restoration with two applications in spatial statistics. *Annals of the Institute of Statistical Mathematics*, 43:1–59, 1991.

[3] N. Breslow and D. Clayton. Approximate inference in generalised linear mixed models. *Journal of the American Statistical Association*, 88:9–25, 1993.

[4] B. P. Carlin and T. A. Louis. *Bayes and Empirical Bayes Methods for Data Analysis*. Chapman and Hall, London, 1996.

[5] D. G. Clayton. A Monte Carlo method for Bayesian inference in frailty models. *Biometrics*, 47:467–485, 1991.

[6] M. K. Cowles and B. P. Carlin. Markov chain Monte Carlo convergence diagnostics: A comparative review. *Journal of the American Statistical Association*, 91:883–904, 1996.

[7] J. Cuzick and R. Edwards. Spatial clustering for inhomogeneous populations(with discussion). *Journal of the Royal Statistical Society,B*, 52:73–104, 1990.

[8] P. Diggle and A. Chetwynd. Second-order analysis of spatial clustering for inhomogeneous populations. *Biometrics*, 47:1155–1163, 1991.

[9] P. J. Green. Reversible jump MCMC computation and Bayesian model determination. *Biometrika*, 82:711–732, 1995.

[10] H. Kunsch. Intrinsic autoregressions and related models on the two-dimensional lattice. *Biometrika*, 74:517–524, 1987.

[11] A. Lawson. Discussion contribution on The Gibbs sampler and other Markov chain Monte Carlo methods. *Journal of the Royal Statistical Society, Series B*, B 55:61–62, 1993.

[12] A. B. Lawson. Markov chain Monte Carlo methods for putative pollution source problems in environmental epidemiology. *Statistics in Medicine*, 14:2473–2486, 1995.

[13] A. B. Lawson and N. Cressie. Spatial statistical methods for environmental epidemiology. In C. R. Rao and P. K. Sen, editors, *Handbook of Statistics:Bio-Environmental and Public Health Statistics*. 1998. to appear.

Systems

INFORMATION SYSTEMS FOR SURVEILLANCE OF COMMUNICABLE DISEASES

Flahault A[1],Parsons DF[2], Garnerin P[3], Valleron AJ[1].

[1] B3E, INSERM U444, ISARS
Faculté de médecine Saint-Antoine
27, rue Chaligny
75571 Paris Cedex 12, France.
Tel: 44 73 84 68
Fax: 44 73 84 54
e-mail: flahault@b3e.jussieu.fr

[2] New York State Department of Health, USA

[2] Hôpital Cantonal de Genève, Suisse

Abstract : Before the Internet era, and prior to the use of teleinformatics, information was gathered via paper-based forms. In this early period, the system of notification of infectious diseases came to maturity and became firmly entrenched in the public health practice of most developed countries. When teleinformatics began to be used in public health, 10 to 15 years ago, the high potential of electronic networking became clear as a result of several national and regional experiments. The New York State Department of Health implemented 'HEALTHCOM', a state-wide health care network, and France developed the 'Reseau SENTINELLES'. For several years, the Centers for Disease Control (CDC) in USA has been alerting state epidemiologists through an electronic text-message system.

Concurrent development and use of relational design of data bases provided appropriate tools for large population data management. In addition, the widespread use of client-server architectures allowed for running of simultaneous applications on the same databases.

In the present era of widespread use of the Internet, it is possible to make the information available worldwide for health professionals, the public, and the media. Since the incidences of diseases contain no patient identifiers, they can be placed in the public domain on the Internet without any security risk.
When the format of notification forms have become internationally standardized in a sufficent number of countries, we will be able to quickly compare the national weekly reports, just by using cross-referenced hypertext pointers on the WEB.

The concept of a rapid, global epidemiological alert system is now emerging, with proposals such as ProMED electronic-mailing lists that enable subscribers to discuss and become informed about any disease. More recently the WHO-FluNet project has provided connections between 110 national influenza-reporting laboratories, allowing for on-line data entry 24hrs a day, and real-time return of processed information. Worldwide spread of influenza strains is now reported as maps, graphs, or tables.

Thus, the management of worldwide public health threats can be managed on the Internet in the near future.

1 Introduction

Collecting, analysing and distributing timeliness information are the main features of information system for surveillance of communicable diseases. Their objectives are to help health authorities to take appropriate prevention measures. Therefore, information systems are closely linked to action, i.e. prevention, in order to control communicable diseases.

In addition, information systems in public health surveillance lead to harmonisation in information processing. They provide materials and tools for early alert detection of outbreaks, or emerging agents, or emerging antimicrobial resistance. Surveillance systems were implemented independently in each country. Recently computers and thereafter the Internet were introduced as a cornerstone in information systems for surveillance and control of communicable diseases.

2 Prior to the Internet

2.1 Before teleinformatics

Far before computer technologies arose, even more prior to the intensive use of teleinformatics, telex, and telefax, information systems were implemented after world war II. It still remains many stigmats of this era in various health sectors where computerisation did not spread.

That was the golden age of paper based bulletins, reports and certificates, such as death certificates. Up to 1994, in United Kingdom, PHLS received about 250,000 notifiable reports from laboratories, among them 60% to 70% were transmitted via classical mail (the rest was received via E-mail).

Royal College of General Practitioners still uses phone calls for collecting data on community acquired diseases, other systems such as the ancestor of the INSERM's *Sentinel* system, which used prior to 1984 post cards for influenza surveillance.

2.2 When teleinformatics arose

Teleinformatics may be defined as computerized tools for information management using a phone line for transmission. This paper will summarized two experiences, one in New-York City, and the French *Sentinel* system. *Healthcom* is the information system on communicable diseases built in New York State. In USA, County Health Departments (CHD) receive directly notifiables paper-based reports from physicians and laboratories. They forward each week the reports to the State Health Department (SHD). However, up to 1996 (Parsons, 1996), sources of data sent to SHD were disparates, including paper forms and various local PC applications. More recently, New York State has implemented a programme aimed to standardise and computerize its information system. SHD sends on a volontary basis reports on notifiables diseases to Centers for Diseases Control (CDC), Atlanta. The federal structure of the USA, as it is in Germany, does not allow the CDC to impose to SHD federal rules for reporting, but only permits them to motivate the network of SHD to participate to an American information system. Currently, all SHDs use the electronic link (i.e. a standardized weekly form to be E-mailed) built by the CDC, named NETSS. Reversely, the CDCs set up PHNET, a tool for early alert and return of information to SHDs (Halperin, 1992). For a long time, it was based on an electronic text-message system, SHDs received earlier than other readers texts of the MMWR. Thereafter, using the Internet graphic capacities, maps and graphs are now available.

The Sentinel system gather about 1% of French general practitionners (i.e. about 500, Figure 1), distributed throughout the continental country (Fourquet, 1997). It uses teleinformatics from 1984. Based on the French Telecom program of spreading its *Minitel*, it allows voluntary and anpaid GPs in theses task, to enter data 24 hrs a day, and at least once a week, according to a specific protocol. Individual reports are collected on influenza, acute diarrhea, chickenpox, measles and mumps, male urethritis, HIV and HCV status. Since July 1997, Sentinel GPs also report hospitalisations they refer, and follow-up information up to 30 days after discharge. Robustness, easyness, and wide availability of the French Minitel was the key of success of a rapid and large adhesion of sentinel GPs in the system.

After 14 years, large databases were constituted, with more than 200,000 cases of influenza-like-illness, individually described. The acquired experience was useful for research purposes, since the time-space series allowed to evaluate mathematical modelling works (Figure 2), which were mainly theoretical before: e.g. study of the dynamic of influenza epidemics (Flahault, 1988). Such fluxes of information stimulate research and development of methods to represent time and space data: e.g. implementation of geographical krigged maps (Carrat, 1992), early warning system of outbreaks detection (Costagliola, 1991 ; Flahault, 1995). The cornerstone of the system is the instantenous return of information, first directed to data providers, i.e. GPs, but also to health authorities and lay public. This feedback is automatised since the beginning of the system, on the Minitel, with maps, graphs and tables, by using the semi-graphic capacities of teleinformatics. With the Internet (Boussard, 1995), an interactive site with high graphic capacities is available without access restriction (Figure 3) (http://www.b3e.jussieu.fr/sentiweb).

2.3 Two milestones on the road to the Internet

The Internet did not appear on our screens and phone or cable lines without having taken advantage of the past experiences. Two technologies were considered as important steps for the future success of the Internet, particularely in the field of the information systems for the surveillance of communicable diseases. First, relational design of database management systems (RDBMS) allows for exploring huge databases constituted with health data from large population. Relational links between tables in RDBMS enhance a rapid access to the data, while saving memory allocation.

Large available marketed RDBMS are now reliable, stable and upgradeable. Standard query language (SQL) is widely used, providing a technical tools for sharing information collected by various teams at various periods.

Second, the client-server architecture, which is used with similar technologies on the Internet, allowed for running of simultaneous applications on the same databases. This computer concept changed the time sharing as perceived by the user, from a waiting period during which queries were sequentially analysed, to a parallel process, where multiple users remotely acceed simultaneously to the same database. Several general practitioners could therefore enter data in the host computer, according to the same epidemiological protocol, or may visit the site, without knowing that CPU was shared between many users.

3 Nowdays: the massive use of the Internet

The French experience with the videtotext home computer (Minitel) did not spread within the European Union, probably because of the lack of standards for common data exchange. The answer to most of issues raised during the development of past client-server architectures came with the Internet. It makes information available worldwide to the health professionals, public and the media. As expected, the Internet did not solve all questions which was asked in the recent past. In addition, other issues arose, regarding various aspects: technological, ethical or legal, or scientific.

3.1 Technical issues about the Internet in 1997

This section hopefully will be dated very soon. But we wanted to underline the current limitations of the Internet. The main one is the slowness of the system. Connexion to a site is time consuming, and the site may often be out of real interest, leading to continue the search to other site, where any query may take time. « Pipe-breaks » are of frequent occurrence, closing prematurely the user's session, without any information or caution message. Many French users are disapointed to use a system also much more difficult to install than the old Minitel, which was a « plug and go » home computer. Transmission problems are of particular acuteness when the used line has a low bandwith such as the classical phone line. Faster networks, such as RNIS (numeris in France) are acceptable solution available for the public, but still less powerful than the cable-TV network which does not provide the Internet in countries like France in 1997. We will focus herein on issues related to access to database, and various current proposals to solve the problems.

Common Gateway Interfaces (CGI), were the first method used for database exploring from a WEB browser. These codes are written in Perl, C or C++ on the host server. They are still considered as standards for read or write access to data. They act as follows : the client which is at a remote site sends a query from an HTML WEB page (e.g. asking for a map of the French incidence of chickenpox as diagnosed by general practitioners in May 1997). This query is then addressed to the host server with the parameters defined by the client. CGI is launched at the server site, and it results in a HTML page which is sent back to the sender (i.e. the client's browser). Any query generates a CGI, each of them are sequentially launched at the server site, overcrowding the host computer, particularely when the site is becoming popular and visited. Timeliness of the site may be considerably affected by its success. This method is particularely subject to pipe-breaks, and other transmission weaknesses. From the computer scientist's point of view, coding CGIs is a cumbersome task, each application needing a particular CGI program, leading to multiple similar codes and risks of errors. Most of programs are just copied and pasted with minor changes for running into similar applications. However, CGIs is a reliable technology when trained and skilled computer scientists follow quality assurance principles. Most of current sites allowing to database access in 1997 were written with CGIs. In addition, newer versions, such as fast-CGIs could overcome most of the limitations of CGIs in the next future.

Technics based on *Java Data Base Connector (JDBC)* may be among the most appropriate solution for the future approach. Codes are written in Java which is a particularely adapted language for the new information technologies. Java has many advantages in comparison to the CGIs. Servlets take place of CGI, by dealing with the part of queries which need to be desserved by the host server. In the Java philosophy, this section reduced to the maximum, leaving most of the work to the client, where applets perform data analyses. In our previous illustration, the graphical and statistical analyses of data are done at the client micro-computer, by using an applet which is included in an HTML page. Connexion with the host server is always short periods, and data exchanges are as light as possible, without any transmission of heavy images, or sounds, or video formats. All these added values are produced at the client site, whenever it is possible. The server site becomes lightened, and may deliver its tasks more rapidely. In addition, servlets allow for real parallel exploration of the database from several clients. From the computer scientist point of view, Java coding is associated with the concept of class, which are blocks to be reused for other applications: double codes are avoided. Applications coded in Java are plateform independent (may be installed either on a Unix machine or a PC or a Macintosh for example).

3.2 Regulatory issues

Since they became popular in the media, issues regarding privacy or security may appear to be more acute with the Internet, but they were not specific. These issues arose earlier. By contrast, we may consider the fact that the Internet is probably the architecture which is taking advantage of a global attention as no other client-server architectures has benefited before. Many proposals for implementing privacy protections and security measures are now available, and will appear in the next years for the Internet users.

Privacy protection in France is submitted to a specific law which limits drastically all attempts to link files or records. Experience regarding sharing medical records by using the Internet technologies are rare in France. However these tools are probably currently the appropriate ones for networking community care with hospital environment or other health ressources (e.g. laboratory investigations, imaging).

Exchanging epidemiologic information between States does not imply any confidentiality issues, since totally anonymous data are involved, such as incidences or prevalences. For maintaining a control on information spread, access codes and passwords may restrict the access to authorized users.

Security measures are sometimes more complex to implement, since any system may be hacked. The aim is to complicate or delayed the intrusion. Guidelines are now available (in France, since June 1997), for recommending measures to adopt in case of an application involving medical data which are accessible within a network. One of the main principle is to separate machines which store nominative data, with those which store other medical data. This separation may be a transcient link which can be randomly activated for establishing links between data when necessary (e.g. updating the database). Another principle is to limit to authorized computers the access from outside into the network of computers. This limitation can be done by using a firewall, which is a specific computer which its software. Accredited firewalls have proven their efficacy.

3.3 Scientific debatted issues

The use of the Internet is particularely increasingbetween scientists. New questions arose, in particular in the field of public health surveillance and action, and in multicentric research.

In the specific domain of communicable diseases, the use of the Internet leads to a higher level of harmonisation between surveillance systems and States, or international organisations. Standards are to be defined regarding exchange formats, but also time periods of exchange, or amount of information to be shared. Moreover, although not specific to the Internet use, but enhanced by it, use of a common language, with common definitions of medical or epidemiological terms are of great importance. For instance, flu, flu-like-illness, influenza, sore-throat, upper respiratory viral infections may represent distinct nosological entities within a same country, or may have different meanings between countries. With the need to exchange and share information at a global scale, in order to measure pandemic phenomenon, or take appropriate common action, consensus on terminology is more and more crucial to obtain. States will not accept to take costly measures from data collected abroad with a debatted level of accuracy, or with large flaws in definitions used. ICD-10 does not answer to many aspects of primary care, and that leads to develop specific classification, such as the international classification for primary care (ICPC).

A terme, mais probablement avec davantage de réticences de la part de certains Etats, les échanges internationaux suscités par Internet, notamment dans le domaine des maladies transmissibles, conduiront à une harmonisation du choix des mesures de prévention.

The final goal which may be achieved because of the Internet, particularely in the field of communicable diseases is to harmonize strategies of prevention. However many countries may have reluctance on that point. Quite recently, but a little prior to the Internet era, the Global concern on AIDS led to adopt or recommend prevention measures, and also to reject unappropriate measures, such as the immigration restrictions. States have been criticized for their health policies, considered as conterproductive action. The enhanced access to information will most probably be accompanied with an increase in this kind of international ban, particularly with regard to health policies.

4 Early global alert systems are arising

4.1 ProMED (program for monitoring emerging diseases)

One of the pionneer, promoted by the non profit federation of American scientists, ProMED is a network, built in 1994 with its communication arm *ProMED-mail* which is actually active (Morse, 1996). It is an open list of E-mail users registered as experts from their countries, allowing for early detection of epidemics of any nature or origin, and even prior to their confirmation. This open forum between participants is aimed to spread widely information and new ideas, at a scale and a speed never reached before. Between 1994 and 1996, more than 300 outbreaks have been reported through this system. The main limitation of this system is that it is not conceived for systematic collect, analyse and distribution of information, since neither the tool nor users skills allow for these missions. It is then hard to expect that ProMED will become a system for diseases surveillance in the future (Lancet, editorial, 1996).

4.2 The European project IDA-HCSSCD

This system is devoted for data exchange between regulatory authorities in charge of control of communicable diseases. Funded by the European Union in 1996, it was aimed to distribute to authorized users (through securized access) available database, such as those concerning legionnella, salmonella or AIDS. The pilot phase was announced in late 1997.

4.3 WHO projects

The EMC division in WHO, Geneva, developed a global monitoring electronic system for Influenza (FluNet), in collaboration with INSERM Unit 444. FluNet provide a remote securized access through the Internet to 110 national influenza centres and 4 collaborating centres. Data entry on a standardized electronic form is available, and a real-time return of information provide maps (Figure 4) and graphs (Figure 5) with the space and time dynamic of the epidemics at a global scale. The time step for data collection is a weekly basis. Interactive queries are available to explore the database, which will be distributed in 1998 to the WEB public without any restriction.

Other similar projects are in progress according to the similar methodologies. One is about human and animal rabies (RabNet), the other one regards the global monitoring of antimicrobial resistance (ARMnet). Both should work routinely in 1998.

5 Conclusion

It is time now to manage and control major epidemic threats by using new information technologies, such as the Internet.

These technologies allow for collecting data in real time from any remote site in the world, for distributing the analysed information immediately to the public, using multimedia interfaces making the message easy to understand. Health authorities may take timeliness appropriate decisions, with the help of international experts and organizations.

Most efforts should be focused in attempting to harmonize information systems, which remain highly heterogeneous either in quality and quantity of their deliverables. The needs for an international harmonization in this field appear to be the first step before sharing complete and detailed informations between countries.

6 References

AHCPR (Agency for Health Care Policy and Research). Moving toward international standards in primary care informatics: clinical vocabulary, October 1996 ;69.

BOUSSARD E, FLAHAULT A, VIBERT J-F, VALLERON A-J. Sentiweb: French communicable disease surveillance on the World Wide Web. Br Med J, 1996; 313:1381-1384.

CARRAT F, VALLERON A-J. Epidemiologic Mapping Using the "Kriging" Method. Application to an Influenza-Like Illness epidemic in France. Am J Epidemiol, 1992; 135:1293-1300.

COSTAGLIOLA D, FLAHAULT A, GALINEC D, GARNERIN P, MENARES J, VALLERON A-J. A routine tool for detection and assessment of epidemics of influenza-like syndrome in France. Am J Public Health, 1991; 81:97-99.

FAST TRACKING EPIDEMIC INFORMATION (editorial). Lancet, 1996; 348: 1599.

FLAHAULT A, LETRAIT S, BLIN P, *et coll.* Modelling the 1985 influenza epidemic in France. Statistics in Medicine, 1988; **7**, 1147-55.

FLAHAULT A, GARNERIN P, CHAUVIN P, *et coll.*Sentinelle traces of an epidemic of acute gastroenteritis in France. Lancet, 1995; 346:162-163.

FOURQUET F, DRUCKER J.Communicable disease surveillance: the Sentinel network. Lancet, 1997; 349:794-795.

HALPERIN W, BAKER EL, MONSON RR. Public health surveillance. Van Nostrand Reinhold, 1992 (238 pages).

MORSE SS, ROSENBERG BH, WOODALL J, et coll. ProMED global monitoring of emerging diseases: design for a demonstration program. Health Policy, 1996;135-153.

PARSONS DF, GARNERIN P, FLAHAULT A, GOTHAM I. The status of electronic reporting of notifiable conditions in USA and France. Telemedecine J, 1996; 2:273-284.

Figure 1 : Geographical distribution of Sentinel General Practitioners (SGPs) in France in 1997 (each dot represents one SGP) - 1% of all GPs are included in the sample.

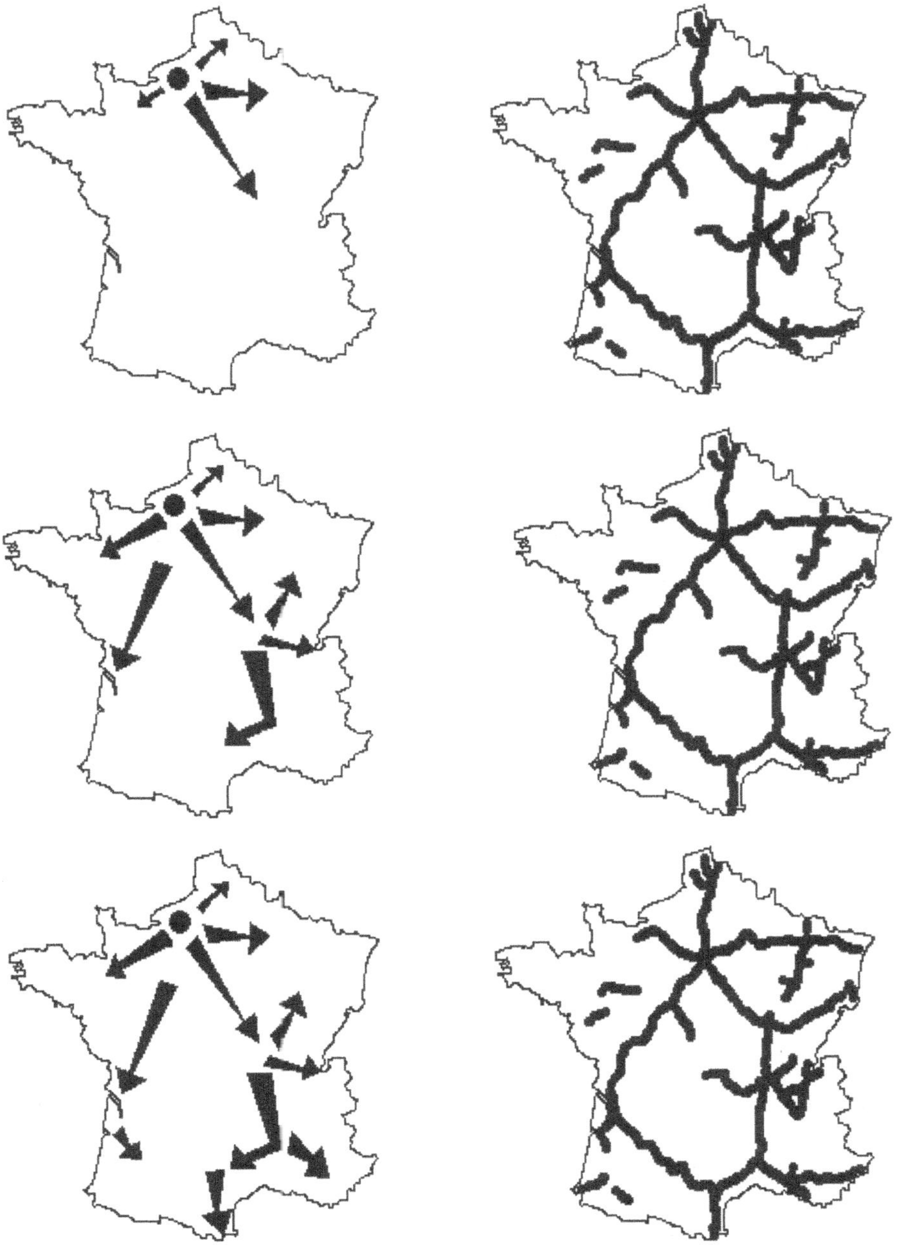

Figure 2 : Comparisons between the time-space dynamic of an Influenza epidemic as observed from the Sentinel system (arrows represent the weekly geographical dynamic of the epidemic on maps at the left side) and motor highways in France (network on maps at the right side).

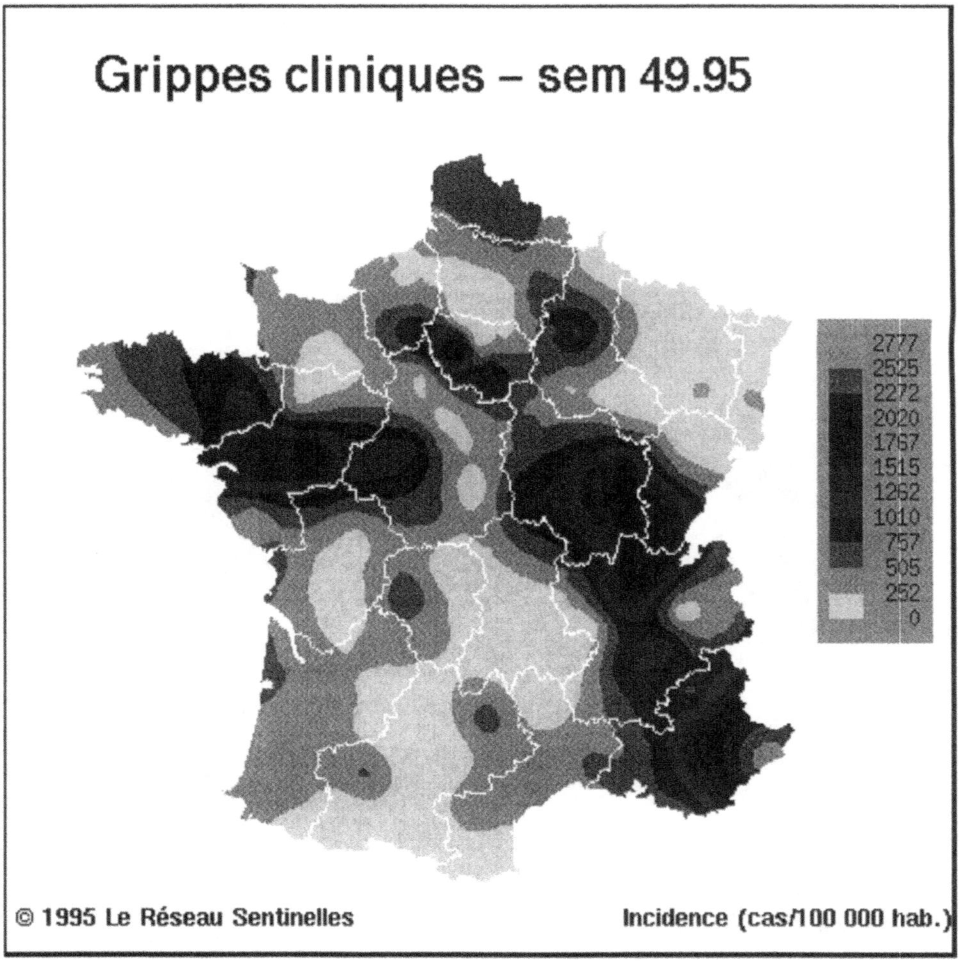

Figure 3 : On the fly output provided by the *sentiweb* site (URL : http://www.b3e.jussieu.fr/sentiweb). This map is obtained just by clicking with the mouse on a friendly user interface for selecting the time period and the disease of interest.

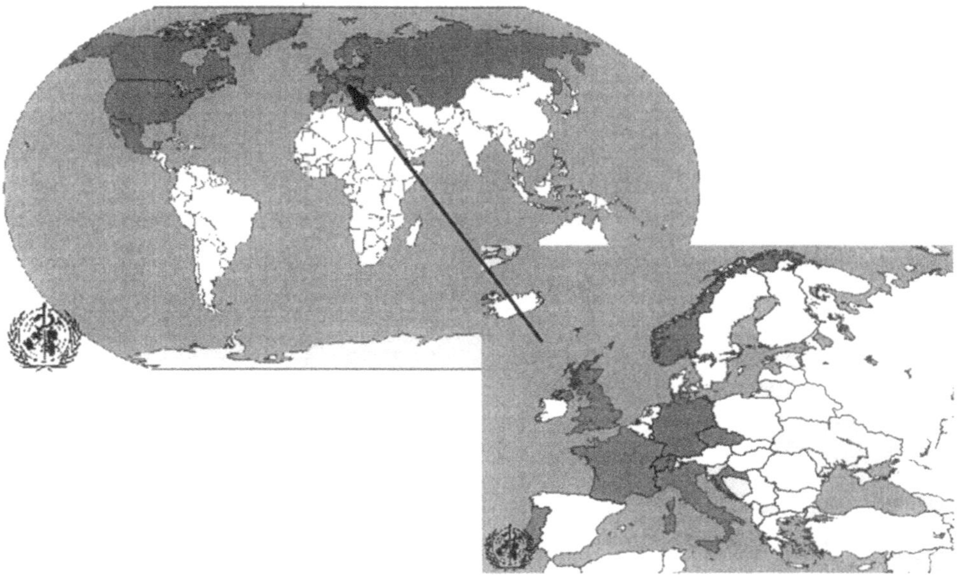

Figure 4 : FluNet : Global electronic monitoring of Influenza, available on the web since Jan, 1997 (URL : http://www.b3e.jussieu.fr/flunet). After remote data entry, a real-time access to updated data allows for displaying maps or graphs. By clicking into a continent in the worldwide map, corresponding countries appear on the screen, and more precise data are also available. (With the permission of WHO, EMC, Geneva, Dr Daniel Lavanchy - WHO is not responsible for this display)

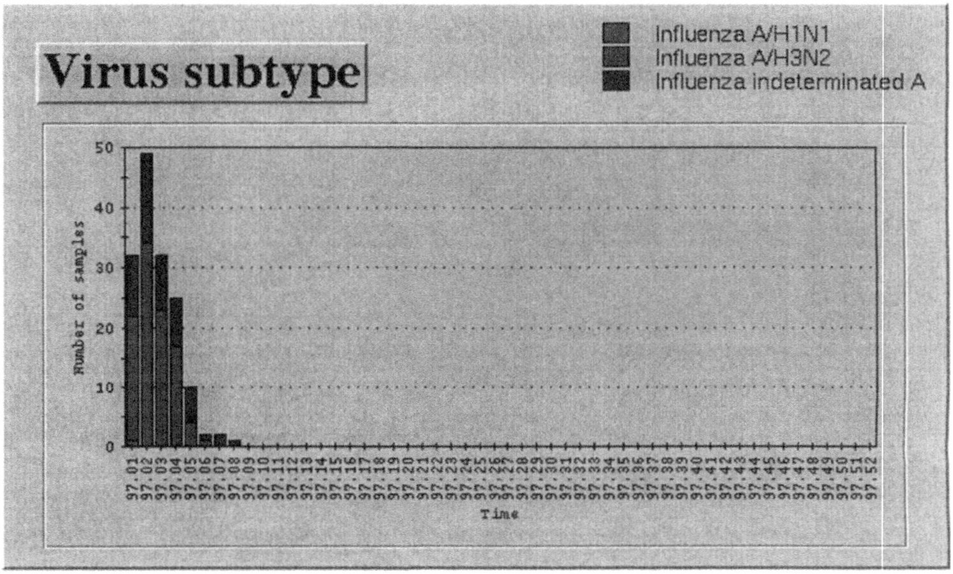

Figure 5 : FluNet graphs display bar charts with Influenza isolates. Trends may be monitored in a specific country, continent or worldwide. (With the permission of WHO, EMC, Geneva, Dr Daniel Lavanchy - WHO is not responsible for this display)

The European Influenza Surveillance Scheme
First Experiences With An Internet Application

Helmut Uphoff
Arbeitsgemeinschaft Influenza AGI
Schuhmarkt 4
D-35037 Marburg
E-mail: helmut.uphoff@kilian.de

Abstract

The European Influenza Surveillance Scheme was founded in 1995 and included seven countries (Belgium, France, Germany, Netherlands, Portugal, Spain, and the United Kingdom). The Czech Republic and Switzerland will join in the season 1997/98.
The roots and hence experience of the group can be traced back to the Eurosentinel-(1987-91) and ENS-Care-project (1992-93), later the early warning system-Influenza until 1994.
The main objectives of the EISS-group are to provide actual information on the influenza situation in Europe in the sense of an early warning function, during the weeks that influenza epidemics are likely in this area, research in epidemiology, economy and virology of influenza as well as the comparability of the national data.
The weekly electronical data transmission as well as the presentation of the results is done by an internet application at the moment. The internet application has been developed on the basis of the scientific work that has been done regarding quality of the included data, comparability of the national data and experiences with previous computer based applications. During the next season pages for public access will be added containing a mapped overview of the situation in Europe as well as graphs of the clinical and virological data in each country including the national comments.

Introduction

Influenza is a major public health problem especially with regard to the growing proportion of elderly and other people at high risk [17,19,22].
The huge financial burden due to a high morbidity in the entire population may exceed billions of ECU in the EU during epidemic years [25].

Almost each winter a phase of elevated influenza-activity can be clearly identified by virus isolations and morbidity [3,16,22]. The intensity and impact of those epidemics vary considerably depending on the prevalent influenza-type, -subtype and even -variant, the affected age groups, the immunity of the population and

many other factors as indicated for instance by the seasonality [5,12]. Nevertheless a substantial number of excess hospitalisations and deaths especially among the risk groups is caused by influenza outbreaks almost every winter [7,23].
This situation could be changed and worsened dramatically if major drift or a shift, the total exchange of the surface antigenes haemagglutinin and or neuraminidase by reassortment, would occur [18].

The available antiviral substances amantadin and rimantadin can be used prophylactically or therapeutically but has to be applicated shortly after first symptoms occurred. They are only active against influenza A and the use is mainly limited by side effects. The possible occurrence of resistant strains has to be considered as well [1,4].

Vaccination has proved to be an efficient and cost-beneficial measure in reducing morbidity, complications, and death. Due to the declining titre of the induced protective antibodies and the occurrence of antigenic drift it should be performed yearly especially by the people at risk. Protection depends mainly on the match of the circulating viruses with the vaccine viruses [7,11,12,21].

The GP based influenza monitoring systems are close to the community and very efficient especially when supplemented by virus isolations from the observed population. Those surveillance systems allow the detection of sporadic influenza activity several weeks before the beginning of epidemics. First local outbreaks that usually indicate the beginning of the epidemic can be spotted on time. An early identification of the age groups most severely affected by influenza is possible as well as -unusual clinical manifestations or complications. First signals for the dynamic and impact of the influenza activity can be analysed about three weeks after the beginning and the approach of the peak is usually indicated one week before it is reached [2,3,6,10,13,26]. But the exact when - and where - the epidemics start and what intensity and impact of the influenza activity should be expected can not be predicted [15,22,23].
In addition to early warning, the value of the monitoring lies in the assessment of the risk of the population or sub-populations explaining the value of vaccination (and other protective measures if available) in order to raise the coverage of the risk group which still is rather low in many countries [14].
A high number of virus isolates from the population enhances the identification of the prevalent virus types and variants as well as the early recognition and evaluation of importance of new drift variants (or shifts). This finally supports the

world wide collection of isolates by the WHO which is necessary for the identification of the variants with importance for the upcoming season [3,12,22].

As the quality of the national data has a major influence on any further evaluation and research, EISS formulated minimum requirements. All of the participating national monitoring systems collect clinical as well as virological data on a weekly base. The virological and clinical data should be linked and focused on the same population sample. The availability of the data of the reported week is requested to be no later than Wednesday the following week. The surveilled population - possibly by means of random samples - as well as the geographic expansion of each system should be nation wide. A certain stability of the national systems and continuous work for at least two successive observation periods should be given. The surveilled period should include the span from October to April.

Additional data regarding age groups, certain regions, symptoms, antibody assays, prevalence of other agents etc. can be included if available.

Due to the remaining differences of the national monitoring systems and healthcare-systems the comparability of the national data is still a problem.

The direct cooperation of the epidemiologists and virologists with responsibility for the national monitoring within the EISS group enables a good assessment of the quality of the data.

As a fundamental instrument for research a joint database containing virological and clinical data of the participating countries was established.

Materials and Methods

Internet application:

During the observation period 1996/97 the internet application of the EISS-group was established. The application was ready to be used from week 46 in 1996 but as not all members had internet connection at that time, Belgium, France and Germany started and were joined by the other members. During this first test phase only the public link page was accessable without authorization. All the other menues as shown in Fig.1 were protected.

„Links“ were offered - to the pages of some organizations (e.g.WHO) and institutions - as well as mailing lists. Under „Users“ the homepages of the institutions or organizations conducting the national influenza surveillance could be found besides the E-mail addresses of the EISS-members. The „Download“ functions allowed an easy downloading of the data of the current week or the complete season in several formats. The „Live“ page provided the opportunity of database record querries.

Data entry:
Separate pages for the entry of virological and clinical data were offered. The virological data entry for a certain week allowed the separation of the results of swab examinations and serologies and distinguished between samples from sentinel practices and other sources (Fig.2). The number of tested specimens in each of the four subgroups (cultures and serologies from sentinel practices or others) were given optionally. The judgement of the local virologist regarding the dominant type(s) or subtype(s) was obligate. The comment offers the possibility to give further explanations or information, e.g., the match of the epidemic variants with the vaccine or announcements of important or unusual occurrences for example the detection of drift variants.
The entry page for clinical data offered several numerators and denominators (Fig.3). This is necessary because mainly due to differences in the health care systems of the participating nations neither of these parameters are defined uniformly.
Data regarding age groups and sex were supplied optionally. The influenza activity was given in five categories, based on the nomenclature adopted by the CDC: no activity, sporadic-, local-, regional-, widespread-. This information is obligate and used for the mapped overview. Additional information was provided with the comments.
The data entry of the reported week should be done no later than Wednesday the following week. The data was stored in the database and currently entered records for a certain year and week replaced the former record for the respective week.
After processing the data, a mapped overview of the situation could be viewed under the menu „Results“ (Fig.4). A mouse-click on a chosen country showed the clinical and virological data of the selected country as graphs (Fig .5,6).

Results

During the 1996/97 season the EISS application reflected the current morbidity and influenza activity in the participating nations. Especially the map based on the epidemiological and virological comments gave an up to date overview and was easy to understand.
The presented results of the clinical data differ regarding the shapes, proportions and scales of the graphs and do not allow to compare the intensities and impact of the influenza epidemics between countries. The course of the epidemic in time (start, peak, end) however can be compared and provide valuable information together with the comments of the experts who know the structure of the concerned national healthcare systems and the validity and reliability of his data.

The virological data do not always reliably reflect the influenza activity within the respective nation and does not allow to compare the influenza activity between nations as well.
Nevertheless, the predominant influenza type and subtype, based on the data from primary care practices was always indicated in the current comment for each county. The graphs of influenza virus detection showed a good relation to the clinical data in all systems and allowed the linkage of elevated morbidity to the activity of one or more influenza types and subtypes. Important events like first detection of an other influenza type or subtype were mentioned and assessed in the comment.
Together with the numerical virological and clinical data important information regarding influenza activity (no activity, sporadic-, local-, regional-, wide spread-) and the predominant type(s) or subtype(s) were presented with a high actuality by the application.

Similar to the previous years EISS data for the 1996/97 influenza season showed considerable differences in the dates of first virus isolations, time of the beginning of the epidemic and peak of influenza activity. The relative intensity (measured by national systems in relation to previous seasons) of the influenza activity varied as well.
The time lag of several weeks between countries for the beginning of the epidemics as well as the peak activity underline the potential and importance of an European early warning. A reliable European early warning system for influenza is of particular importance in case of an antigenic shift.

Discussion

The continuously updated comments of the participating national epidemiologists and virologists presented by EISS as a mapped overview via internet provided a reliable and topical overview of the influenza activity in Europe during the 1996/97 season. Based on the clinical data and comment the spatial expansion of influenza activity could be estimated. The virological information identified the predominant influenza types and subtypes. The added comments of the experts that are in charge of the national influenza monitoring systems had already taken into account the known limitations of the national data and reduce misinterpretations.

The clinical data represented in the graphs using different numerators, denominators and hence scales are not easy to interpret and provide no comparability. During epidemics designated as moderate in different countries for

instance the observed values can be ten times the background value in one country or less than twice in another. Besides systematic differences caused by the chosen numerator and denominator, the measured morbidity is additionally influenced by mechanisms that determine the interactions between GP and population on a national and or regional level. Interpreting the data is possible only for people with profound knowledge in epidemiology and the respective national health care and monitoring system.
Therefore, based on several years of experience of the group, work is in progress to develop an influenza activity index based on the relation of the current situation to past epidemics for each country. This first step enables to give an easy to read impression of the current influenza activity in relation to „normal" or „unusually high" in each country. This approach does not consider Possible „long term" differences in „true" influenza activity between countries throughout the years or systematic differences (like different sensitivities due to different numerators). The harmonisation of the data may also help to harmonise the terminology describing the intensity of influenza activity as „normal winter activity", „elevated-„ or „epidemic activity".

The data regarding the influenza virus isolations and antigen detections allow the link between elevated morbidity and influenza activity of a certain type or subtype [2,12,24]. For each country, the indication of the dominating influenza virus (sub)type(s) is based on the isolates obtained from patients sampled by GPs because the source of the examined swabs is critical to interpret the data. Hospitalized patients constitute a small highly selected group for which e.g. the severity of the disease will in general be greater. Consequently the ratios of the various (sub)types found in hospitalized patients may differ considerably from those found with patients of GPs [8]. The virological data of the current two weeks are preliminary and need updating because the determination of the subtypes of influenza A positive samples require successful virus isolation which may take up to ten days [22,23]. Due to differences in the national virological monitoring comparability of the virological data is not sufficient. Further standardisation in origin of the samples (selection of patients) and laboratory procedures would be a major enhancement to the EISS system and will be soon on the agenda.

The influenza activity in the different countries varied considerably in time and relative intensity for this observation period. This has been observed frequently especially during inter-pandemic periods [22,23]. No obvious pattern in the spread of virus or in the dynamics of influenza activity (time wise and space wise) could be inferred from the data collected over the recent years. Research regarding the

epidemiology and virology of influenza has to be intensified. In this respect it might be helpful to break down national data in order to find geographical zones in Europe which experience at the same time over a wide range of years.
The present joint database of the EISS including most of the European countries with influenza monitoring systems that match the required minimal standards is an excellent fundament for such research. Besides collecting and presenting the current data EISS will advance research aspects in order to increase the quality and reliability of early warning.
Early warning has to be managed cautiously and ist value mainly depends on ist reliability. Institutions, organizations and persons in key positions for the preparation for influenza outbreaks will „learn" to react adequately only when the early warning is reliable. Therefore the EISS group decided to lay emphasis on the quality of national data, stepwise harmonization in cooperation with the national experts and research regarding validity, predictive value and interpretation of the data. For the time being, the internet pages of EISS will present only the current influenza situation in Europe and will refrain from giving any predictions.
International exchange of data for ongoing epidemics is important for many diseases and influenza might be a pioneer.

Acknowledgements

This paper has been written on behalf of the EISS members: Aymard M (Fr), Bartelds AIM (NI), Cloetta J (CH), Cohen JM (Fr), de Jong JC (NI), Fleming DM (UK), Havlickova M (CR), Manuguerra JC (Fr), Marinho Falcao I (P), de Mateos S (E), Mosnier A (Fr), Noone A (UK), Perez-Brena P (E), Rebelo de Andrade H (P), Snacken R (B), Tumova B (CR), Valette M (Fr), van der Velden J (NI), van der Werf S (Fr), Watson J (UK), Yane F (B) and Zambon M (UK).

Literature

1) Anonymous: Prevention and Control of Influenza: Part II, Antiviral Agents Recommendations of the Advisory Committee on Immunization Practices (ACIP). MMWR 1994;43 RR-15
2) Bartelds AIM: Zur Validierung von Sentinel Daten. Das Gesundheitswesen 1993;55:3-7
3) Bartelds AIM, Fracheboud J, Van der Zee J, eds. The Dutch sentinel practice network; relevance for public health policy. Studies from the continuous morbidity registration in The Netherlands. Utrecht: Netherlands Institute of Primary Health Care (NIVEL), 1989
4) Belshe RB, Burk B, Newman F, Cerruti RL, Sim IS: Resistance of Influenza A Virus to Amantadine and Rimantadine: Result of one Decade of Surveillance. J Infect Dis 1989;159:430 - 435
5) Beveridge WIB. Influenza the last great plague. Heinemann educational books 1977

6) Birkhead G, Ghorha TL, Root S, Klaucke DN, Gibbs NJ. Timeliness of national reporting of communicable diseases: the experience of the national electronic telecommunications system for surveillance. Am J Publ Health 1991; 81: 1313-1315
7) Carrat F, Valleron AJ: Influenza mortality among the elderly in France, 1980-90: how many deaths may have been avoided through vaccination?. J. Epid and Community Health 1995;49:419-425
8) Claas ECJ, de Jong JC, Bartelds AIM, Bijlsma K, Rothbart P, de Groot R, Rimmelzwaan GF, Osterhaus ADME. Influenza types and patient population. Lancet 1995;346:180
9) Colacino JM. Resistance of influenza A and B viruses to 4-guanidino-Neu5Ac2en. Options for the control of influenza III 1996; Elsevier Science: 741-747
10) Costagliola D: When is the epidemic warning cut-off point exceeded? Eur J Epidemiol 1994; 10: 475-477
11) Couch RB, Keitel WA, Cate TR, Quarles JA, Taber LA, Glezen WP. Prevention of influenza virus infections by current inactivated influenza virus vaccines. Options for the control of influenza III 1996; Elsevier Science: 97-106
12) Cox NJ, Regnery HL. Global influenza surveillance: tracking a moving target in a rapidly changing world. Options for the control of influenza III 1996; Elsevier Science: 591-598
13) Dab W, Cohen J M, Hannoun C: A new influenza surveillance system in France: The ile-de-France "Grog". 2. Validity of Indicators. Eur J Epidemiol 1991;7(6):579-587
14) Fedson DS, et al : Influenza vaccination in 22 developed countries: an update to 1995. Vaccine (England) 1997; 15(14):1506-15011
15) Fleming DM: The prediction of epidemics of respiratory infections. Eur J Epidemiol, 1994; 10: 481-484 7th European Meeting of Influenza and its Prevention
16) Fleming DM, Cohen J M: Experience of European collaboration in influenza surveillance in the winter of 1993 - 1994. J of Public Health Med 1996; 18: 133-142
17) Glezen WP. Serious morbidity and mortality associated with influenza epidemics. Epidemiol Rev 1982; 4: 25-44
18) Hampson AW, Cox NJ. Global surveillance for pandemic influenza: Are we prepared? Options for the control of influenza III 1996; Elsevier Science: 50-59
19) Itzstein von M. Design and synthesie of influenza virus sialidase inhibitors: Novel anti-influenza drugs. Options for the control of Influenza III 1996; Elsevier Science; 697-701
20) Kim C U, et al.: Influenza Neuraminidase Inhibitors Possessing a Novel Hydrophobic Interaction in the Enzyme Active Site: Design, Synthesis, and Structural Analysis of Carbocyclic Sialic Acid Analogues with Potent Anti-Influenza Activity. J Am Chem Soc, 119 (4), 681-690
21) Levy E. A cost benefit study of flu vaccination for 25- to 64-year-old French people. Options for the control of influenza II 1993; Excerpta Medica: 47-52
22) Snacken R, Lion J,van Casteren V, et al.: Five years of sentinel surveillance of acute respiratory infections (1985-1990) the benefits of an influenza early warning system. Eur J Epidemiol 1992; 8: 485-490
23) Snacken R, Bensadon M, Strauss A: The CARE Telematic Network for the Surveillance of Influenza in Europe. Meth Inform Med 1995;34:518-522
24) Sprenger MJW, Beyer WEP, Kempen BM, Mulder PGH, Masurel N: Risk factors for Influenza mortality?. Options for the control of Influenza II; Excerpta Medica 1993:15-23
25) Szucs TD. The economic impact of influenza in Europe. Options for the control of influenza III 1996; Elsevier Science: 34-42
26) Van Casteren V, Leurquin P. Eurosentinel: Development of an international sentinel network of general practitioners. Meth Inform Med 1992; 31: 147-52

ropean Influenza Surveillance Scheme - Internet Database http://www.eiss.org/db/main.htm

European Influenza Surveillance Scheme

Results
Consult the latest weekly compiled on-line graphics.

Live
Review the latest 'on-line' results.

Data Entry
Send through your institution's results.

Download
Select from our file list of statistics in various formats.

Links
Useful links to other systems and also the Influenza Mailing Lists.

Users
Contact points for data contributors and sponsors of the EISS web site.

For technical correspondence only
System design copyright ©, QUAD LOGIC 1996
Last updated 15 Nov, 1996

Figure 1

uropean Influenza Surveillance Scheme - ArbeitsGemeinschaft Influenza http://www.eiss.org/db/agi/dbentryV.htm

Menu | Results | Download | Links

European Influenza Surveillance Scheme

Virological Data Entry for Germany ArbeitsGemeinschaft Influenza

Your results should be keyed into the fields provided in the 'Virological Data' section. Once complete choose the *'Submit Data'* button to send through your results. Should any errors be made, modify the data and then select *'Submit Data'* again. The data compilation will be based on your last results filed for that specific year / week combination. The *'Clear Data'* button will erase the form and allow you to re-key.

Region : Germany
Year :
Week :

Virological Data

Sentinel

Number of Specimens	A Not Sub-typed	A (H_1N_1)	A (H_3N_2)	B	C	RS
Number of Serologies	Serologies A	Serologies B	Serologies RSV			

Others

Number of Specimens	A Not Sub-typed	A (H_1N_1)	A (H_3N_2)	B	C	RS
Number of Serologies	Serologies A	Serologies B	Serologies RSV			

Dominant Type, Subtype

None
Type A *Sub Type H_1N_1* AND OR *Sub Type H_3N_2*
Type B

Comments

Figure 2

ropean Influenza Surveillance Scheme - ArbeitsGemeinschaft Influenza http://www.eiss.org/db/agi/dbentryC.htm

Menu | Results | Download | Links

European Influenza Surveillance Scheme

Clinical Data Entry for
Germany ArbeitsGemeinschaft Influenza

Your results should be keyed into the fields provided in the 'Clinical Data' section. Once complete choose the ***'Submit Data'*** button to send through your results. Should any errors be made, modify the data and then select ***'Submit Data'*** again. The data compilation will be based on your last results filed for that specific year / week combination. The ***'Clear Data'*** button will erase the form and allow you to re-key.

Region : Germany
Year :
Week :

Clinical Data - Germany

Results

Number of ILI *(Numerator 1)*

	Age 0-4	Age 5-14	Age 15-64	Age 65+	Unknown
Male					
Female					
Unknown					

	Age 0-4	Age 5-14	Age 15-64	Age 65+	Unknown
Population *(Denominator 2)*					
OR					
Number of Encounters *(Denominator 2')*					
OR					
Number of Physicians *(Denominator 2")*					

Activity

No Activity | Sporadic | Local Outbreak
Regional Outbreak | Widespread Outbreak

Comments

Figure 3

n Influenza Surveillance Scheme - Statistical Analysis http://www.eiss.org/db/consult.htr

Menu | Live | Data Entry | Download

European Influenza Surveillance Scheme

Statistical Analysis of Data
Latest Compilation

You may view graphical charts for any of the participating countries. Please select from the map below:-

Please click on a country for more detail.

Belgium | France North | France South | Germany | England | Scotland | Wales | Netherlands | Portugal | Spain

For technical correspondence only
System design copyright ©, QUAD LOGIC 1996

Figure 4

Menu Data Entry Download Links

European Influenza Surveillance Scheme

Graphical Analysis of Data for the Netherlands

The results shown below are based on data from the first recorded week upto the latest received results.

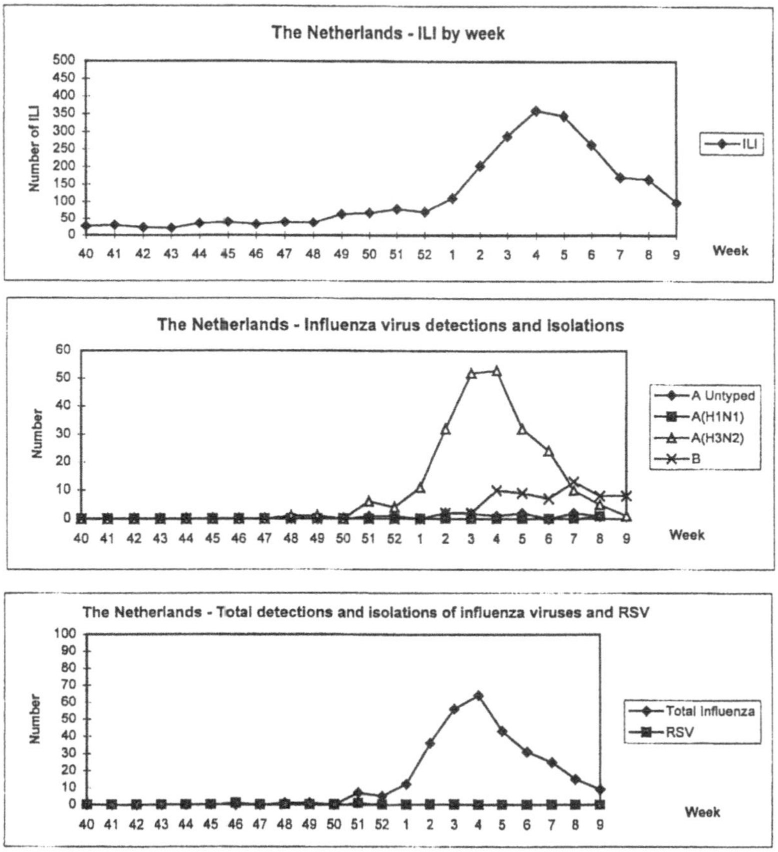

Figure 5

European Influenza Surveillance Scheme

Graphical Analysis of Data for Germany

The results shown below are based on data from the first recorded week upto the latest received results.

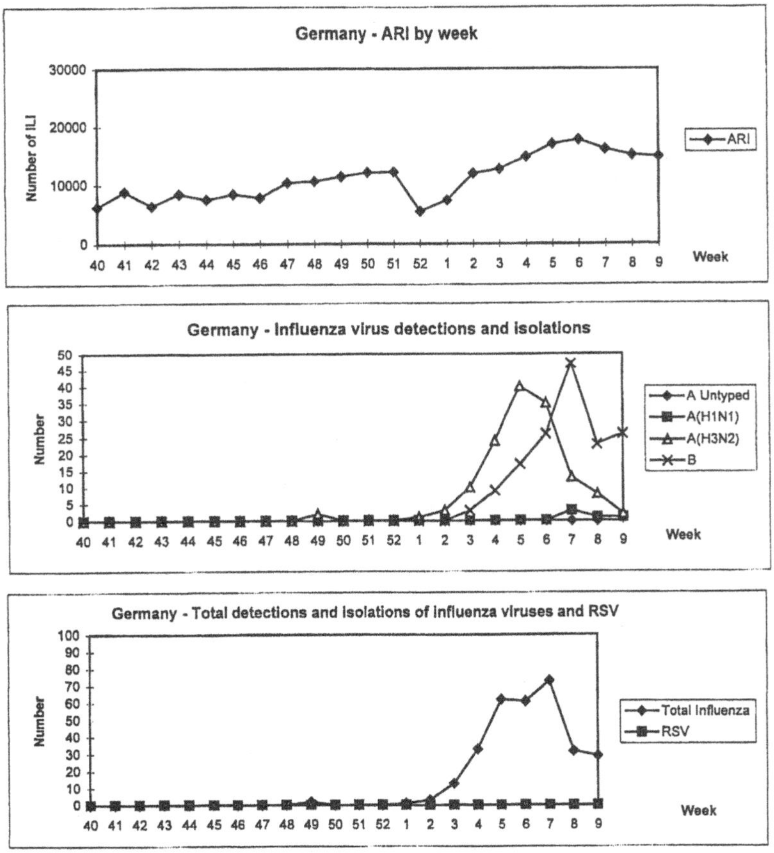

Figure 6

Spatial Data Analysis Support for Cancer Epidemiology in *CARESS*

Frank Wietek[1], Vera Kamp[2]

[1]Oldenburger Forschungs- und Entwicklungsinstitut für Informatik-Werkzeuge und -Systeme (OFFIS), Escherweg 2, D-26121 Oldenburg,
e-mail: wietek@offis.uni-oldenburg.de

[2]Universität Oldenburg, FB Informatik, Escherweg 2, D-26121 Oldenburg,
e-mail: kamp@informatik.uni-oldenburg.de

Abstract

CARTools, a toolbox designed to support population-based cancer registries, is currently being developed at OFFIS, an institute for computer science. The first application area of this toolbox is the registry being established in Lower-Saxony, a federal state of Germany. *CARLOS* (Cancer Registry Lower-Saxony) is the name of the corresponding project, started in 1993. The *CARTools* comprise four tools, one of which is *CARESS* (*CARLOS* Epidemiological and Statistical Data Exploration System), implementing various routines for statistical and epidemiological data analysis on the registry database.

In this paper, the functionality of *CARESS* with respect to supporting spatial data analysis – which is the main focus of this system – and the design of the integration of data analysis features in an interactive graphical user interface will be presented. The system is supposed to provide different features and interfaces for different user groups (epidemiologists, registry staff, external researchers) and different tasks (e.g. quality control, incidence monitoring, cluster analysis). To ensure both efficiency and flexibility, the underlying database system strives to integrate concepts from the fields of spatio-temporal, multidimensional and statistical database systems.

1 The Project *CARLOS*

To reinforce efforts in the fight against cancer and to improve knowledge about potential causes and risk factors for cancer – aiming at an improved prevention –, the Ministry of Social Affairs forces the establishment of a homogeneous population-based cancer registry in Lower-Saxony, a federal state of Germany, since 1992. The name of the corresponding project, which started in 1993 with a pilot period, followed by a test period 1995-97, and which is presently beginning on routine data collection, is *CARLOS* (Cancer Registry Lower-Saxony) [AFH+96].

Evaluation of the applicability of a model for cancer registration proposed by Prof. Michaelis (Mainz) [MK92], also the Federal Cancer Registration Act passed in 1995 is based on, has been the main task in the pilot period. This model suggests a distinction between two separate offices: A notification office („Vertrauensstelle“) collects and encrypts patient-related reports, before transferring them to the registration office („Registerstelle“), which persistently stores, links together, and analyses the data. Concepts for data encryption and record-linkage have been designed, evaluated, and standardised [AMIT96].

Apart from OFFIS, which will take over the part of the registration office, the main collaborating partners in the project are – under the management of the ministry – the Association of Panel Doctors in Lower-Saxony („Kassenärztliche Vereinigung Niedersachsen (KVN)“) and a number of additional cancer-registering institutions, like clinical registries, pathological laboratories, etc. The KVN is expected to contribute the largest amount of reports to the registry by way of the „Nachsorgeleitstellen“, which are institutions engaged in patient follow-up and organisation of secondary therapy and further examinations.

Currently, the project is in the final stage of a test period lasting from 1995 to 1997. During this time, special emphasis has been and is put on the integration of a variety of cancer-registering institutions, the implementation of routine registration, and computer-based support for epidemiological data analysis.

To provide software support for almost all steps of population-based cancer registration, especially in Lower-Saxony, a toolbox called *CARTools* is being developed at OFFIS [AFH+96]. The *CARTools* comprise four tools providing support for the main tasks „data transfer to the registry“, „encryption of data sets“, „record linkage“ and „data analysis“. *CARESS* (*CARLOS* Epidemiological and Statistical Data Exploration System) is the name of the tool implementing the statistical and epidemiological data analysis process based on the registry database.

2 Analysing the Registry Database with *CARESS*

Experiences in a number of cancer registries have shown that there is still a lack of facilities to analyse the large amounts of collected data flexibly and efficiently both for routine monitoring and reporting and to answer public ad-hoc queries regarding supposed clusters of cases or increased incidence rates. Often, data storage and management, statistical analysis and visualisation of results are implemented by different systems and tools, so that data transfer and conversion takes a lot of time or is at least inconvenient. Furthermore, in many cases new problems and queries with just some parameters changed afford new software routines to be written and integrated.

The data analysis system *CARESS* is designed to meet these needs and to make work at population-based cancer registries more efficient and flexible. It is supposed to provide a comfortable data analysis environment for different groups of users (cf. fig. 1) interested in

- quality control, i.e. calculation of indices like HV (proportion of histologically verified tumours), M/I (relation of mortality to incidence), DCO (number of cases only known by means of a death certificate), etc.,
- incidence monitoring, survival analysis, and cluster analysis,
- formation of hypotheses concerning potential cancer risks by comparison with background information,
- generation of reports and data export.

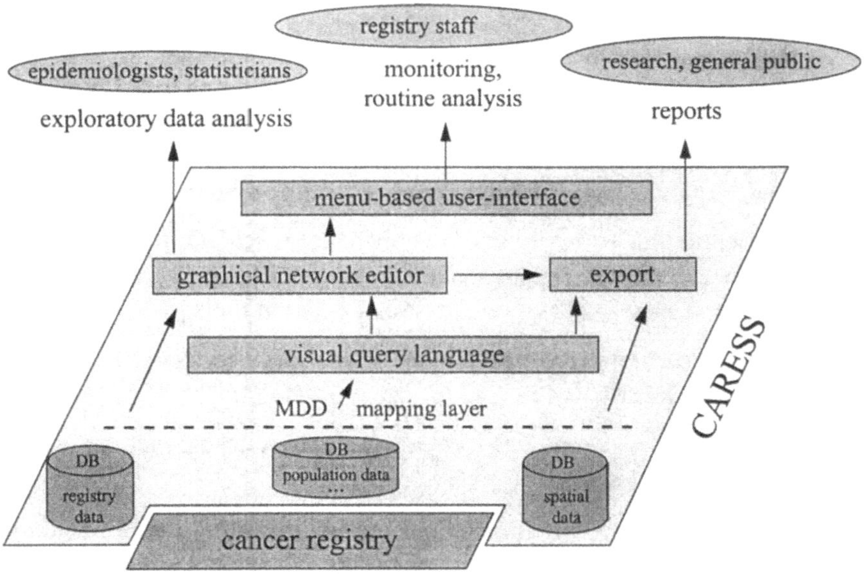

Fig. 1. System architecture and user groups of *CARESS*

We designed the system not to be restricted to utilisation in the registry of Lower-Saxony, but to be portable at least to other population-based cancer registries and even to be applicable in similar domains like health reporting or descriptive epidemiology in general.[1] Thus we chose the metaphor of an „epidemiological

[1] Since 1996, the cancer registry of Hamburg also evaluates *CARESS* in analysing the Hamburg cancer database.

workbench", being modular and extensible with respect to new methods for analysis or visualisation and interfaces to further tools and systems.

Figure 1 outlines the architecture of *CARESS*. Tumour- and patient-related registry data is linked together with population data and additional spatial data into a unified view onto the database. This view is implemented by a MDD (multidimensional discrete data) mapping layer, which is based on concepts developed for spatio-temporal, multidimensional and statistical database systems [Bau94, CCS93, Gue94, Sho82]. Aggregated data sets (e.g. case or population counts by region, age, and sex) are implemented as data cubes, providing efficient access to groups of single data values. By defining categorisation hierarchies on classifying attributes (space, time, age, sex, or type of disease), flexible aggregation of data in different granularities is supported. At present, we make use of the relational database management system ORACLE, while also evaluating object-oriented and object-relational concepts.

To keep the user from having to formulate complex SQL-queries to select and combine data, we plan to implement a visual query language based on the MDD mapping layer mentioned above. This language should comprise statistical, spatial and temporal operators as well as data management facilities for multidimensional data, especially aggregation. Further components of the system architecture are

- a graphical editor to easily combine different calculation and visualisation procedures, to compare algorithms and data sets within an exploratory data analysis, and even facilitate integration of external tools [WK97],
- a tool to define and execute sequences of analysis steps for report generation and data export, and
- a menu-based user interface providing an easy-to-use access to routine calculations and graphics.

Menu-based user interface and graphical editor are discussed in more detail in the following section. Support for report generation is still subject to future work.

2.1 User Interfaces

A number of epidemiological measures (number of cases, crude, directly or indirectly standardised and cumulative rates, SMR, relative risk and so on) can be calculated in *CARESS* for a study population described by different parameters, like place of residence, time, age, sex, kind of tumour, and further medical attributes. Visualisation of results in bar charts and line charts, tables, and above all thematic maps is easily parametrised and interactively modifiable by kind of measure displayed, standard population, type of cases (incidence or mortality), and different specific parameters. Figure 2 gives an impression of the user interface of *CARESS* – in this case an example of a thematic map created by the system. The open menu shows different types of maps provided (grouping by quantiles, maps based on a scale with 3-10 classes of equal ranges, probability maps, etc.). An online help system explains utilisation of the tool, describes all calculations and visualisation procedures, and gives the exact formulas of all measures.[2]

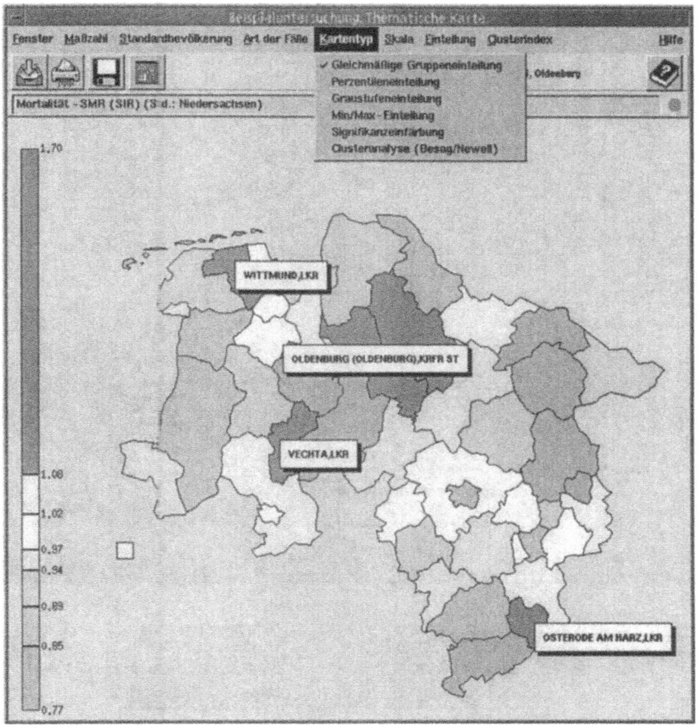

Fig. 2. Example of an evaluation

In the graphical network editor, a study comprising groups of evaluations is modelled as a network of methods. Each node represents the application of one method of a certain method class to one or more data sets, which may be processed in parallel (independently) or in combination as a joint input of one

[2] As this system documentation is implemented using HTML, it is also accessible via WWW: „http://elbe.offis.uni-oldenburg.de:3999/caress-doku/".

calculation. The connections between nodes represent the data flow between data sources, data analysis and visualisation procedures.

Menu-based interface and graphical editor are not meant to be isolated from each other, but the first can be regarded as a special-purpose view onto the second. Each menu selection can be translated into a network of single analysis steps (outlined in figure 3), which may be modified and extended in the network editor for further, more detailed calculations. Thus, both of the user interfaces are based on a single concept of data access and analysis.

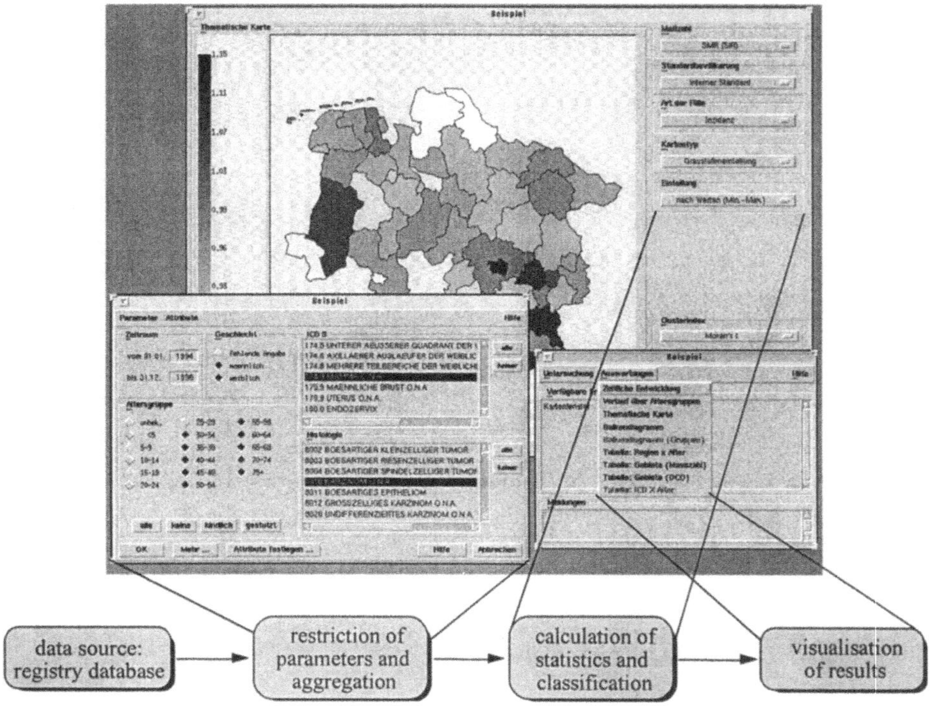

Fig. 3. Translation of menu-based parametrisations to analysis networks

2.2 Basic Model for Data Analysis

Our application model is based on several layers describing a cycle of interactive data exploration (see figure 4). As already mentioned above, an integrated view onto the different kinds of data, especially the combination of statistical and spatial data, constitutes the core of our system. Selection of data requires facilities to perform visual queries and the existence of especially spatial and temporal operators, which may be used in long interactive and iterative data selections. Techniques applied in analysing the data are determined by visual combination of predefined, above all statistical operators, based on aggregation and summation of data. Complex operations can be constructed and combined interactively using concepts of visual programming languages. Additionally, facilities to reuse parts of selected and analysed data in the course of the same study, but also in further data explorations have to be provided. Finally, visual presentation of results both leads to data interpretation and triggers new steps of analysis.

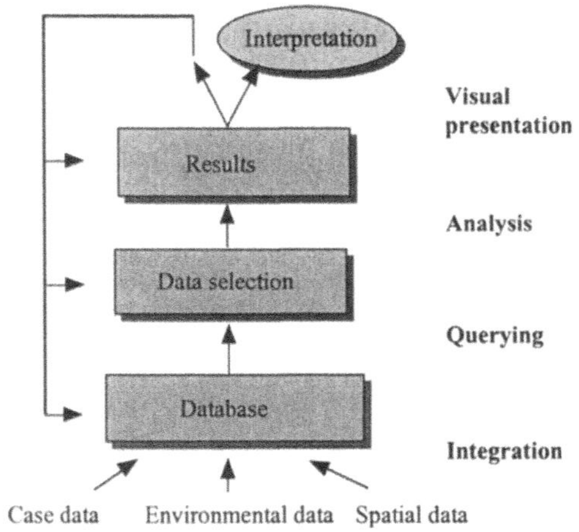

Fig. 4. Interactive data exploration

3 Spatial Statistics

The main focus of *CARESS* is on spatial data analysis. In collaboration with the cancer registry of Hamburg, different measures describing spatial and space-time clustering of cancer cases are being implemented in *CARESS* (cf. [BN91, Wal92]). Work in this area is based on results of the symposium „Methods of Spatial Description and Analysis of Cancer Registry Data", which has taken place in March 1996 in Hamburg [BSMH96]. Cluster indices are supposed to enhance the information presented visually in thematic maps (cf. section 2) – we consider them another way of data description, *not* primarily some kind of significance test.

One can distinguish between global and local cluster analysis, i.e. judging clustering in the region under study as a whole and examining exact locations of clustering. The global cluster indices implemented in *CARESS* can be divided into three groups:

- Moran's I (1948), Geary's c (1954), the statistic of Ohno & Aoki (1979), and the D-test (1985) test for spatial autocorrelation, i.e. similar cancer rates in neighbouring regions. They differ in the use of measures (possibly incl.

standardisation for age, sex, or population size in general), ranks and classifications.

- The test statistic of Potthoff-Wittinghill (1966) and similar indices consider heterogeneity of cancer rates in general or concentration of high rates in few regions without taking neighbourhood of regions into account.
- The test defined by Knox (1963) focuses on space-time interaction: Are cases which are „close“ in space also „close“ in time? This type of test is especially relevant for infectious diseases.

As far as possible, we calculate p-values for cluster indices both using simulations and approximations based on estimates for mean and variance, offering the chance to compare and better understand characteristics of these measures. Decision support, which index to use in which situation, and an improved understanding of how to interpret differences in the results provided by different indices will be subject to further research.

In contrast to the cluster indices listed above, a procedure proposed by Besag & Newell in 1991 helps to find the locations of clustering. The algorithm can be outlined as follows:

- Select a cluster size $M = k \times$ average number of cases.
- For each region test for clustering around this region:
  - Aggregate neighbouring regions, until sum of cases $\geq M$.
  - Test for significance w.r.t. aggregated population based on a Poisson distribution of cases and an expected number of cases derived form the whole population under study.
- Highlight „significant“ regions.

Complementing the objective of „improving“ thematic maps regarding their power to judge clustering, we are also starting to evaluate, design, and implement procedures for smoothing cancer maps (distance-weighted smoothing, empirical Bayes estimators, splines etc.) in our project.

4 Integration of Geographical Data

The most important spatial data source of *CARESS* is *ATKIS* („Amtliches Topographisch-Kartographisches Informationssystem" – Official Topographical Cartographic Information System), a database built up by the federal survey-offices [AdV91]. *ATKIS* provides – beneath data describing the borders of administrative units – a large number of geographical objects from the basic German map („Deutsche Grundkarte"), classified in different groups (e.g. streets, power plants, high voltage mains).

The project *InterGIS* [Fri97] aims at defining and implementing a flexible interface to *ATKIS* via Internet. *CARLOS* or *CARESS*, respectively, takes the place of a client of the geographical data services defined by the geographical data server. This also includes a library of geographical operations (like *along of*, *around*, etc.) located at the client (see figure 5). These operations serve as the basic tool for selecting the respective population under study and for correlating cancer rates with background information associated with geographical objects, thus guiding the formation of hypotheses concerning potential risk factors for cancer. These „first ideas" may be subject to further investigations in adjacent case-control or cohort studies.

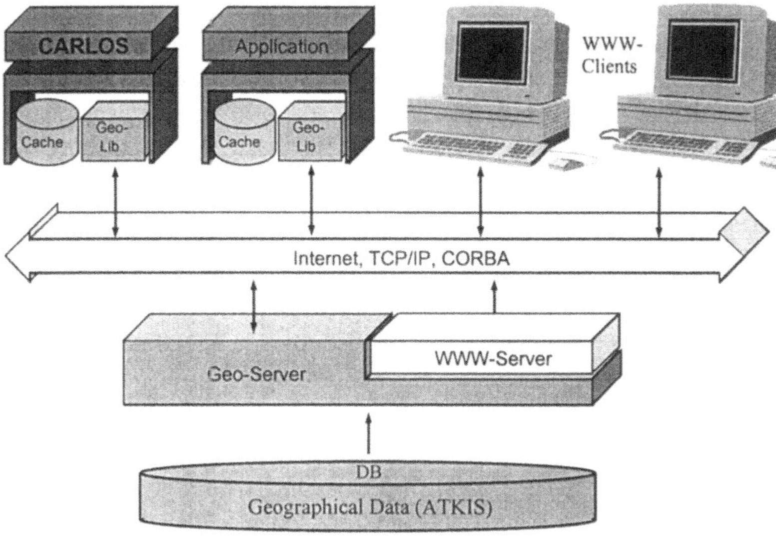

Fig. 5. Architecture of a geographical data server

5 Summary and Future Work

So far, menu-based user interface and graphical network editor are isolated tools, with the editor being in a prototypical stage and the menu-based system already being employed in first routine data analysis. The next step will be to integrate these two systems into a homogeneous environment. Besides, in collaboration with the cancer registry of Hamburg, the system is continuously evaluated, enhanced by new routines for data analysis, and improved with respect to user-friendliness and efficiency. New concepts for data management, especially of spatial and multidimensional data, are evaluated and integrated.

We are aware of the problem known as „fishing for significance". *CARESS* is a powerful data analysis tool that encourages the user to use and compare a variety of algorithms with similar aims (e.g. cluster analysis) on the same data set or even on a number of data sets describing different subsets of the population under study. A system of this kind may also be misused by users, who don't take into account the danger of misinterpreting statistically significant results because of ignoring the context of the analysis carried out (or even those who do, but don't care). Although we want to emphasise, that *CARESS* is just a tool, which provides data analysis facilities and the usage of which the user himself is responsible for, we strive to facilitate a reasonable usage of our system by providing as much information about the methods applied and a sensible way to interpret their results – also in combination with other procedures – in a user-friendly online help system. The integration of concepts from the field of knowledge-based systems, guiding the users in his choice and warning him in case of potentially misleading results, is subject to future work (see [WK97] for some first ideas).

Once *CARESS* will be implemented as a complete data analysis environment in the cancer registry of Lower-Saxony, we aim at transferring the tool to related domains, thus showing the flexibility and extensibility of the underlying basic concepts and system architecture.

References

[AFH+96] H.-J. Appelrath, J. Friebe, H. Hinrichs, V. Kamp, J. Rettig, W. Thoben, and F. Wietek. CARLOS (Cancer Registry Lower-Saxony): Tätigkeitsbericht für den Zeitraum 1.1.-31.12.1996. Technical Report, OFFIS, Oldenburg, 1996.

[AFH+96] H.-J. Appelrath, J. Friebe, H. Hinrichs, V. Kamp, J. Rettig, W. Thoben, and F. Wietek. Softwarewerkzeuge für (epidemiologische) Krebsregister. In: M. P. Baur, editor, *41. Jahrestagung der GMDS,* Bonn, Sept. 1996.

[AMIT96] H.-J. Appelrath, J. Michaelis, I. Schmidtmann, and W. Thoben. Empfehlung an die Bundesländer zur technischen Umsetzung der Verfahrensweisen gemäß Gesetz über Krebsregister (KRG). *Informatik, Biometrie und Epidemiologie in Medizin und Biologie*, 27(2): 101-10, 1996.

[AdV91] Arbeitsgemeinschaft der Vermessungsverwaltungen der Länder der Bundesrepublik Deutschland (AdV). *ATKIS-Gesamtdokumentation, Teil D, ATKIS-Objektartenkatalog.* Niedersächsisches Landesvermessungsamt, Hannover, 1991.

[Bau94] P. Baumann. On the Management of Multidimensional Discrete Data. *VLDB Journal. Special Issue on Spatial Database Systems*, 4(3): 401-444, 1994.

[BSMH96] C. Baumgardt-Elms, M. Schümannn, S. v. Manikowsky, and U. Haartje, editors. *Symposium "Methoden regionalisierter Beschreibung und Analyse von Krebsregisterdaten",* Hamburg, March 1996. Behörde für Arbeit, Gesundheit und Soziales, Temmen Verlag.

[BN91] J. Besag and J. Newell. The Detection of Clusters in Rare Diseases. *Journal of the Royal Statistical Society*, 154(1): 143-155, 1991.

[CCS93] E. F. Codd, S. B. Codd, and C. T. Salley. Providing OLAP (On-line Analytical Processing) to User Analysts: An IT Mandate. White Paper, Arbor Software Corporation, 1993.

[Fri97] J. Friebe. Eine GeoServer-Architektur zur Bereitstellung geographischer Basisdaten im Internet. In: K. R. Dittrich and A. Geppert, editors, *Datenbanksysteme in Büro, Technik und Wissenschaft,* pages 251-60, Ulm, March 1997. Springer Verlag.

[Gue94] H. Güting. Spatial Database Systems. *VLDB Journal*, 3(4), 1994.

[MK92] Michaelis, J., Krtschil, A.: Aufbau des bevölkerungsbezogenen Krebsregisters für Rheinland-Pfalz. *Ärzteblatt Rheinland-Pfalz,* 45(10): 434-438, 1992.

[Sho82] A. Shoshani. Statistical Databases – Characteristics, Problems and Some Solutions. In: *8th International Conference on Very Large Data Bases (VLDB),* pages 208-222, Mexico City, Mexico, 1982. Morgan Kaufmann.

[Wal92] S. D. Walter. The Analysis of Regional Patterns in Health Data. *American Journal of Epidemiology*, 136(6): 730-759, 1992.

[WK97] V. Kamp and F. Wietek. Intelligent Support for Multidimensional Data Analysis in Environmental Epidemiology. In: X. Liu, P. Cohen, and M. Berthold, editors, *Advances in Intelligent Data Analysis - Reasoning about Data. Second International Symposium, IDA-97,* volume 1280 of *Lecture Notes in Computer Science*, pages 299-310, London, August 1997. Springer Verlag.

MONITORING THE PARTICIPATION OF SENTINEL GENERAL PRACTITIONER WITH THE HEALTH CARE WORKSTATION SITIE

Laurent Toubiana and Antoine Flahault

B3E, INSERM U444, ISARS
Faculté de médecine Saint-Antoine
27, rue Chaligny
75571 Paris Cedex 12, France.
Tel: 44 73 84 52
Fax: 44 73 84 62
e-mail: toubiana@b3e.jussieu.fr

Abstract : The Sentinel system was established in 1984 for the national surveillance of communicable diseases. An epidemiological database is maintained and updated practically in real time. Sentinel system collects data from approximately 500 sentinel general practitioners (SGP) spread throughout the country, using computer terminals and more precisely videotex (a low cost telecommunication terminal of the French public telecommunication company).
SITIE is a workstation devoted to epidemiological data selection, analysis and representation. Disease selection, spatial and temporal representations can be done using a user friendly graphic interface. SITIE is built upon the AVS graphics package, and allows one to create new applications using previously developed modules linked together to form a network. SITIE can be viewed as a tool-box for epidemiological data representations.
In this way, we have developed a new tool to follow the activity of sentinel general practitioners using a cartographic representation of their participation.

1 Introduction

The Sentinel system was established in 1984 for the national surveillance of communicable diseases. An epidemiological database is maintained and updated practically in real time. The Sentinel system collects data from approximately 500 sentinel general practitioners (SGP) spread throughout the country, using computer terminals and more precisely videotex (a low cost telecommunication terminal of the French public telecommunication company).
SGP's are asked to report their data at least once a week, even if they do not encounter any case of diseases under surveillance

When such fluxes of data are collected, there is clear needs for quality standards. Standard operating procedures have been implemented either for sentinel practitioners, or for data management. In addition, quality assurance program has to answer each week to the following points :

- how many out of the 500 SGPs reported cases last week ?
- where were they located ?
- were some departments (i.e. regional districts) non represented ?
- was this particular regional outbreak due to an «over- participation» ?

2 Monitoring sentinel general practitioners

SGP's can connect to the computer whenever they want. The distribution of hours and days of connection to the computer shows that SGP's do their epidemiological work throughout seven days per week and almost 24 hours per day.

This freedom in the protocol transmission of epidemiological data enables us to compute the SGP's participation, defined as the number of SGP's by time unit who have contributed to the surveillance.

An SGP is defined as 'active' at time *t* if he has connected to the data centre host at least once during a standard period of time p (in days) which is usually set to 7 days. SGP's are asked to report their data at least once a week, even if they do not encounter any case of diseases under surveillance - in which case they should report 'zero cases'-. This is important in order to discriminate between SGP's who do not see cases and SGP's that are inactive.

To compute the participation we sort all the SGP connections in chronological order during a given time period (most of the time every week). For each connection we compute the time lag from the previous connection. If this time lag is over 12 days (which is the 'exclusion delay'), the data report during this connection is not considered in the incidence. If the time lag is less than 12 days we attribute the value 1 to each days of this period.

For each week and month and for each SGP, we compute the participation, which is the sum of the values divided by the number of days. These computed participations are stocked in 2 data base tables one for weekly participation and one for monthly participation.

3 The health care workstation SITIE

The Sentinel information system is accessed by many users for different purposes. For example, a doctor or a health officer might request statistical informations on a disease in a particular region using different modes of representation; or an epidemiologist researcher might access to the system to provide data for the development of a model. They all utilise the system to extract, treat and represent data which can be performed in many different ways. We observe that users lose time to rebuild general tasks that others have done before. A large amount of epidemiological data has been gathered since 1984, stored in an Oracle ᵗᵐ relational database. At the beginning, data are extracted using hand written SQL requests, stored in files, and processed using specific tools and Mathematica to detect epidemics and represent the spread of diseases. A first attempt to enhance the data processing led to use the couple Oracle-SAS to process the Sentinel system data, using SAS Access to Oracle. However this allowed for a limited user system interaction.

Finally, as an information and research center, the response to users' demands require a comprehensive change. The goal was to develop a system with an integrative architecture which can keep active the old developments and be adapted to new requirements. In order to give a more interactive access to the Sentinel system data base, we developed a system for workstations, structured around the concept of modularity. This workstation offers, in a friendly graphic environment and, in a coherent platform, a set of general procedures to ensure a rapid access to a pertinent epidemiological information.

Non experimented computer users can have an immediate Sentinel system data base visibility without any knowledge on numerous complex softwares. This workstation concept is based on an integration of various tools to selectively extract data, to process and then represent them using either spatial or temporal representations. More advanced computer users can connect these tools graphically to constitute their own developments (applications). Finally C programmers can develop specialised tools that can be used to build new applications.

This software package is called SITIE[1] , a French acronym for *Station Intégrée de Traitement des Informations Epidémiologiques* standing for «integrated workstation for processing of epidemiological informations».

3.1 Concept of SITIE

SITIE is a tool box providing the user with a set of software tools called *applications*, which are full featured procedures easily allowing a complete data processing. An application involves a large number of different tasks: automatic construction of SQL requests, communication with the Sentinel system database, down-loading of the requested data, statistical computations, and graphic representations.

For developing the epidemiological workstation SITIE, we considered a users' needs as a procedure called application. Each application is made of a set of elementary tasks. In fact, we tried to find out the fundamental commons among user's demands. When completed, we obtained common tasks and specific tasks. Actually, we have find out that only few marginal tasks are specific. The great majority of tasks may be applied to multiple contexts in order to build various applications. We have developed the common tasks. Now, it is easy to create numerous new applications by using the common tasks, and by only developing the specific ones.

In fact, a task is designed to be a flexible and powerful processing component, named module. Modules are objects representing the building blocks of an application, they must be linked together to create a complete application. A module

[1]Pronounced "city"

can be viewed as a computational unit. Each of them can accept data as input, compute them and generate data as output. On certain conditions the computing can be modified by passing parameters. Those parameters can be modified by users using a friendly push-button and menu oriented graphic user interface (GUI). From the user side, module are quickly and transparently performed.
Modules can be grouped together according to their function, and archived as modules libraries. It is possible to individualise modules devoted to the data extraction from the data base. Other are devoted to the data processing, such as statistics or filtering, and finally the last group is devoted to the graphic representation of the previous module categories output (histograms, curves, maps, etc.). At the processing chain start, we must found at least one data acquisition module, followed by one or several processing modules, and finally one or several representation modules.

3.2 SITIE architecture

The conceptual idea of module is similar to the concept of subroutine within a structured programming language. Development of modules requires experience in C or FORTRAN programming language.
However, each module can be run individually and be considered as an independent process that have the ability to get information from other modules. It is more like the concept of process in term of system process. Obviously, they can't produce any treatment without others. But when a module is selected, it actually exists and can run a function if it receives a synchronisation signal. The module modifies the data located in a shared memory location. Therefore, the communication between modules is ensured by synchronisation signals and shared memory management.
The main advantage of such a system is to create a new application by adding one or several modules in a preexisting application without a linking pass to get this new application. The only thing to do is to indicate what is the shared memory available and which events may start the process, that can be done graphically. By this way, "intermediate" users can quickly create and check a new application from a previous one. This ensures flexibility and fiability because we do not have to redesign the ancient functionality, we only add a module and check immediately if it works or not.
The next step towards efficiency is the manipulation of modules for building or modifying applications. Each module is symbolically associated with a graphic icon. All valid modules belong to a graphic specialised library where the user can pick up the module he needs and drag it in a working window: the *network editor*. Each module has input(s) and/or output(s), graphically represented by small rectangles whose colour represents a given input or output data type. This facilitates the construction of the application: a given output colour can only be connected to the same input colour. The application building is thus reduced to the module selection and to their connection in order to obtain a *network*. These links correspond symbolically to a data flow, but actually are semaphore gateways in order to operate on data fields located in shared memory. This task may require

some training in building complex applications, but this can be done without any knowledge of advanced programming languages.
Modules and links constituted a network that is a graphical representation of an application. Users can save networks and store them in a application library that can be accessed by the mean of application menus. These menus can be personal or can be access by a large public if its applications are seen like general interest applications.

3.3 Special applications to follow the network activity

In this way, we have developed some new tools to follow the activity of sentinel general practitioners using a cartographic representation of their participation. Figure 1 shows the modules network of the simplest application for the participation monitoring of SGP's and the figure 2 shows the graphical user interface for this application.

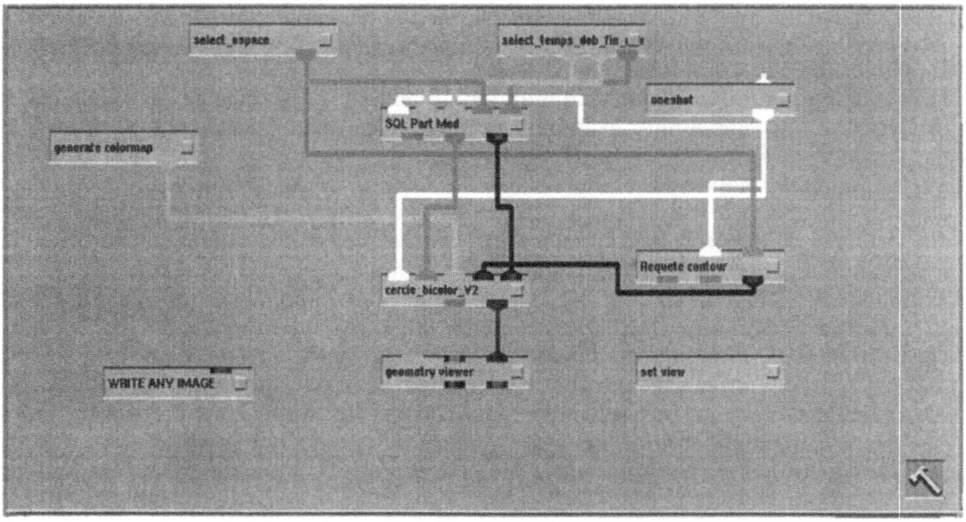

Fig. 1. Modular architecture of SITIE

For the special application, we only develop one specific module SQL_PART_MED which is a generator of SQL requests to the Oracle database. All other modules used in this application had been developed for other previous applications. This is possible because of the structured approach concept of SITIE that allows the utilisation of the same existing modules in several different applications.

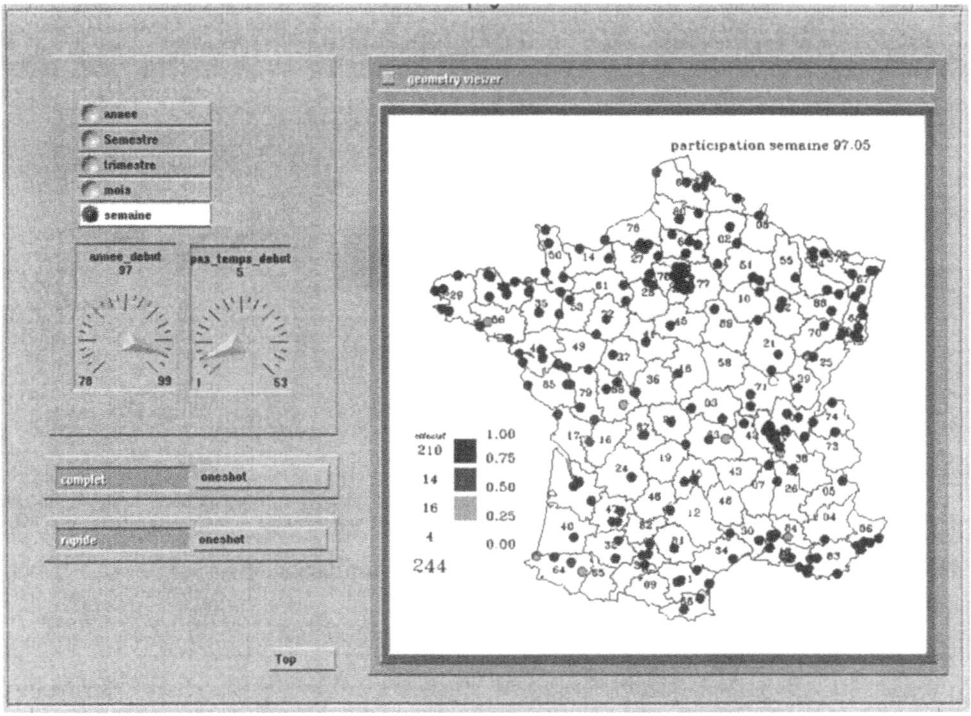

Fig. 2. Graphical user interface of the Sentinel participation monitoring tool

4 Results

4.1 A temporal application to follow the sentinel general practitioners activity

The available tools provide a weekly monitoring bulletin for the purpose of quality control, including a bar chart graph of the daily distribution connexions of Sentinel general practitioners (Figure 3 is a cumulative analysis). It shows that connexions are made at any day of the week, with two peaks (Monday and Saturday). The distribution of connexion per hour is given in Figure 4. It learnt to the monitoring team that physician may be connected to the host server 24hrs a day ; connexions were particularly rare from 2:00 am to 5:00 am.

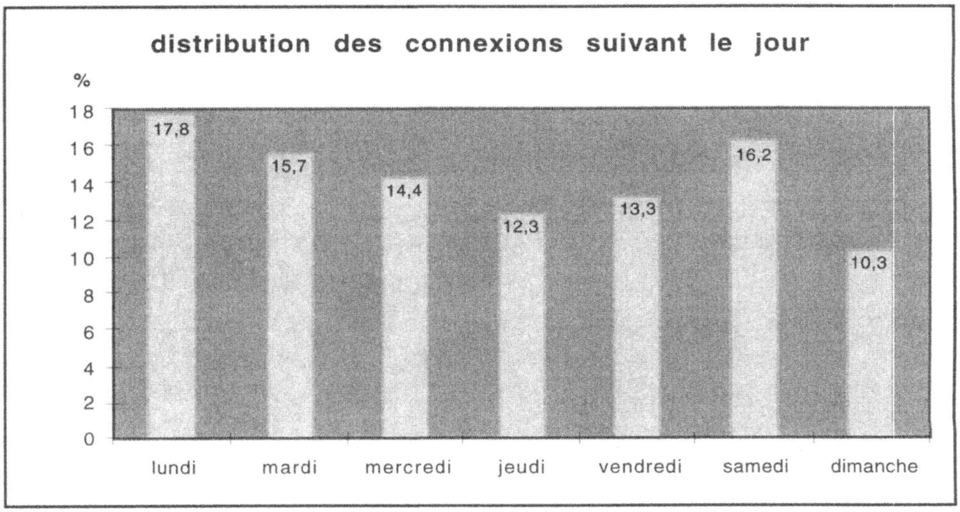

Fig. 3. Output of the Weekly monitoring bulletin - daily distribution of connexions of Sentinel general practitioners

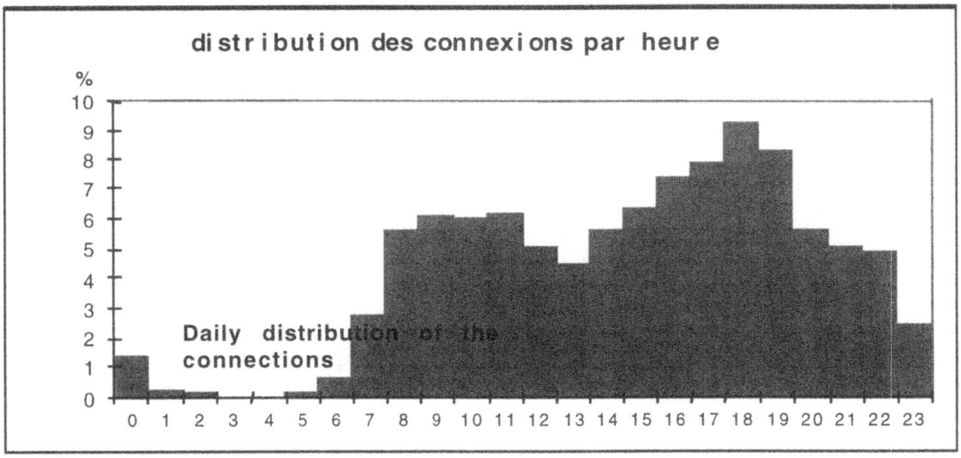

Fig. 4. Output of the Weekly monitoring bulletin - distribution per hour of connexions of Sentinel general practitioners

4.2 Spatial applications

4.2.1 A spatial application to follow the sentinel general practitioners activity

To allow the team in charge of the monitoring activities, SITIE provides maps for targeting SGPs to be sent reminders. For a given week (Figure 5), the color of circle indicates the level of participation and its localisation (figures in the map are the departments codes). Figures at the left side of the legend gives the number of participating general practitioners for a given week. When no sentinel was recorded in a department, reminder is sent to reactivate the network.

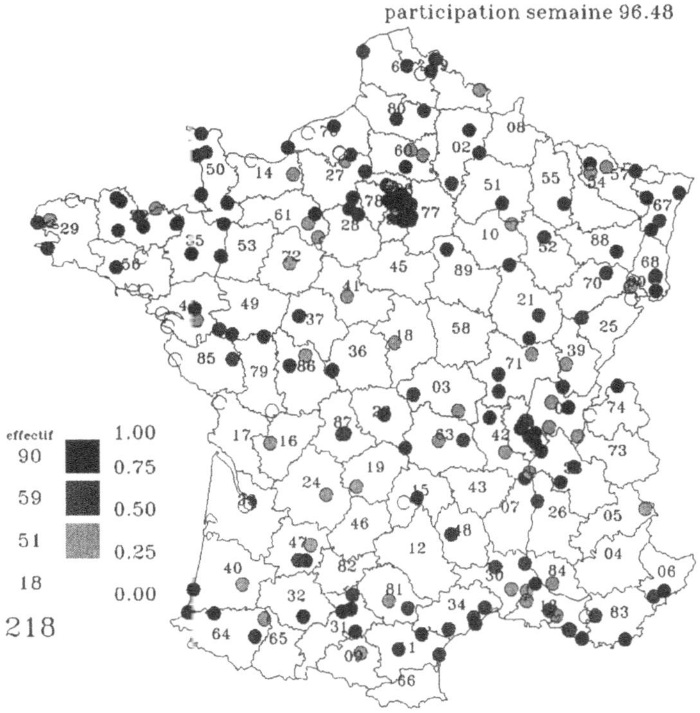

Fig. 5. Output of the Weekly monitoring bulletin - geographical distribution of Sentinel general practitioners in a given week

4.2.2 A spatial application to follow the sentinel general practitioners activity

This tool is not only used for checking the participation of sentinels. It is also useful for verifying if an alert triggered by a rapid increase in incidence of a disease under surveillance is consistent with the participation. For example, it is of high importance to check immediately if such an increase is due to the notification of a unique physician (or a small amount), and therefore to contact notifyiers for confirmation. It is also important to confirm that the absence of increase in other region is not due to the lack of participation. Figure 6 shows the participation map corresponding to the influenza map at the early beginning of a past epidemic. It demonstrated that the focus observed around Lyon was not the fact of an over participation in this region.

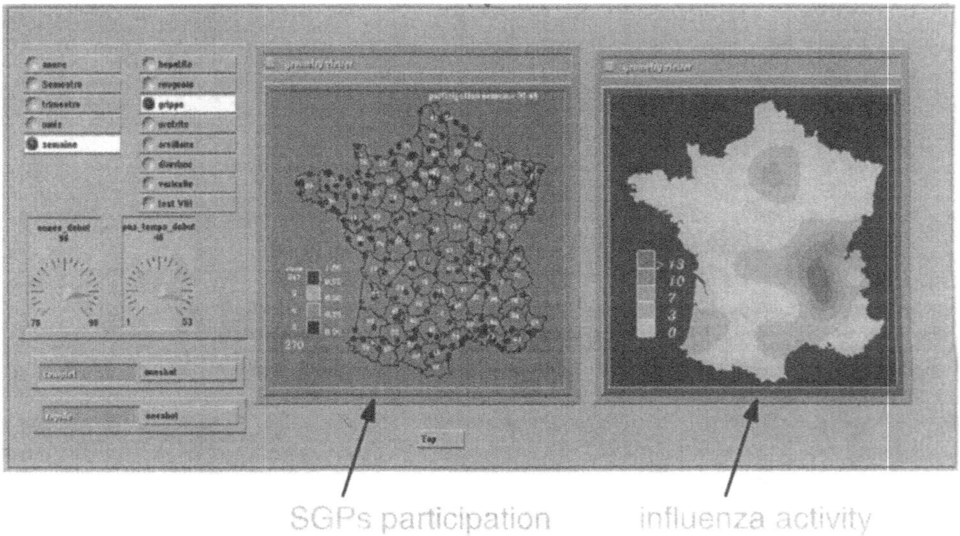

Fig. 6. Graphical user interface of the Sentinel participation monitoring tool. This application shows the participation map corresponding to the influenza epidemic map.

5 Discussion - Conclusion

SITIE is an application developed for exploring epidemiological database with a friendly user interface. It is a tool devoted mainly to researchers interested in the Sentinel system, while Sentiweb (URL : http://www.b3e.jussieu.fr/sentiweb) is mostly public oriented. With its new module, SITIE is also useful for data quality assurance, particularly in monitoring the weekly participation of sentinel general practitioners.

Quality in data acquisition, and analyses is of high importance in epidemiology. When huge databases are managed for surveillance and research in large population (e.g. more than 200,000 cases of influenza are individually described in the Sentinel data base), automated procedures have to be implemented to standardize and facilitate monitoring activities. Other functions of SITIE are under study, regarding automated check of geographical representativeness of the active sample of notifiers, or links between participation and epidemic trends.

6 References

Valleron A-J, Bouvet E., Garnerin Ph., Menares J., Heard I., Letrait S., Lefaucheux J. A computer network for the surveillance of communicable diseases : the French experiment. Am. J. Public Health, 76:1289-1292 (1986)

Toubiana L, Vibert JF, Garnerin P, Valleron AJ SITIE: A health care workstation integration architecture for epidemiologists Comput Biomed Res 1995; 28:100-115

Garnerin Ph., Saidi Y., Valleron A-J. The French Communicable Diseases computer Network : a seven-year experiment. In: Parsons D F, Fleisher C M, Greenes R A, Eds. Extended clinical consulting by hospital computer networks. Annals New York Academy of Sciences, 670: 29-42 (1992)

Valleron A-J, Garnerin Ph. Computer networking as a tool for public health surveillance: The French Experiment. Mmwr, 41: 101-110, supplement (1992)

Garnerin Ph., Valleron A-J. The French Communicable Diseases computer Network : a technical view. Comput. Biol. Med., 22: 189-200 (1992)

A Strategy for Geomedical Surveillance Using the Hawkeye Knowledge Discovery System

Daisy Y. Wong [3] Warren T. Jones [3] Stephen E. Brossette[3]
J. Michael Hardin [2] Stephen A. Moser [1]

Department of Pathology [1]
Department of Biostatistics [2]
Department of Computer and Information Sciences [3]
University of Alabama at Birmingham
Birmingham, AL 35294, U.S.A.
{wong, jones, brossette}@cis.uab.edu, mhardin100@aol.com, moser@uab.edu

Abstract

An approach for the coordination of epidemiologic surveillance within local hospitals as well as across multiple hospital sites, which use the Hawkeye knowledge discovery system, is presented. It is proposed that this coordination be supported by agent-based collaboration technology. This framework meets important criteria recommended by ProMED such as a strongly horizontal decentralized structure.

Keywords: Knowledge Discovery, Surveillance, Epidemiology, CSCW, Collaboration, Interpretation, Intelligent Software Agents

1. Introduction

Surveillance systems are important elements of an effective infection control program. The lack of systems which "recognize and promptly report significant changes and trends" has been cited as a barrier to the successful prevention and control of outbreaks [1]. There is a definite need to make more effective use of existing epidemiologic data. To address this need, an epidemiological surveillance system based on knowledge discovery and data mining techniques called Hawkeye has been developed [2]. These techniques are used to augment traditional statistical methods of data analysis. Hawkeye is currently being applied to hospital infection control surveillance [2, 7]. In this paper, we present a description of the Hawkeye system and a strategy for the coordination of epidemiological surveillance over multiple sites using Hawkeye by applying agent-based collaborative interpreting technology. This strategy would have potential application to the global challenges being addressed by ProMED [3].

2. Hawkeye Knowledge Discovery System

Knowledge discovery is defined as the non-trivial process of identifying valid, novel, potentially useful, and ultimately understandable patterns in data [4]. Patterns of interest for Hawkeye consist of comparisons of temporal histories of association rules and their confidence and support measures of discrete or categorical data. A *frequent set* is a set of zero or more items that co-occur in a data set in more than T records. T, the threshold, is specified in terms of some minimum number of records. A frequent set A has *support*, S(A), which is the number of records in which A occurs in the data set. A special frequent set, called the EmptySet, has zero items. S(EmptySet) is N, the number of records in the data set. The *confidence* of an association rule A=>B is the conditional probability of B given A where A and B are non-intersecting frequent sets in the data. Hence, the confidence of A=>B is equal to S(AUB)/S(A).

Hawkeye's tasks are surveillance and deviation detection. To accomplish these tasks, Hawkeye first discovers all association rules in each data partition. Hawkeye then monitors these association rules over time, using their support and confidence, and detects changes or deviations in their behavior. Upon detection of a significant deviation, it triggers an alert and reports a description of the event to the user for interpretation and verification. Association rules and their confidences over time are stored in a domain *history*.

Hawkeye looks for patterns in the temporal histories of association rules. Because only some patterns are potentially interesting, an *interestingness heuristic* is defined to differentiate between patterns that are and are not potentially interesting. The user is the final judge of whether or not a potentially interesting pattern is truly significant. Ideally, the interestingness heuristic should identify only those patterns as potentially interesting, which the user would identify as interesting, while missing none that the user would identify as interesting. In reality, however, it should be able to identify all patterns that the user would identify as interesting without presenting too many patterns to the user that are not interesting.

The history forms the substrate for pattern discovery. Hawkeye analyzes one partition at a time, updates the history, then determines whether the confidence of an association rule discovered has increased significantly over time using a Chi-Square based comparison. If the difference is deemed significant, then the finding is considered potentially interesting and is presented to the user as an alert. Figure 1 provides an overview diagram of the Hawkeye system's alert discovery process.

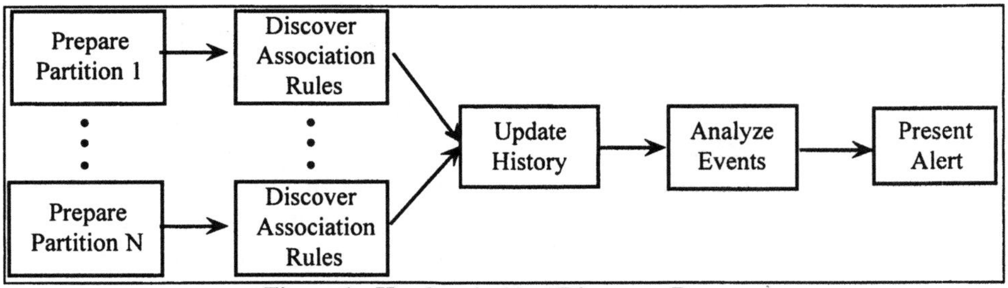

Figure 1 - Hawkeye's Alert Discovery Process

3. Applying Hawkeye in Epidemiologic Surveillance

Hawkeye has been employed to detect the emergence and spread of antimicrobial resistance in a local hospital using its Clinical Microbiology Laboratory Information System data [7]. Sample alerts of *Pseudomonas aeruginosa* susceptibility discovered from data over a period of three quarters in 1996 are shown in figures 2 and 3. Hawkeye's processing and analysis mechanism is described in detail in [7]. In summary, Hawkeye computes the confidence of an association rule for each data partitions that the rule is identified. It then uses a Chi-square test of two proportion to compare the confidence of each rule in the latest partition with confidences from previous partitions. If the Chi-square test shows that the probability of the difference between the proportions occurred by chance is less than five percent, then the association rule is reported as an alert. In Figure 2, the alert shows the detection of an increased significance of resistance to the antibiotics pair Ticarcillin/Clavulanic acid (R~Ticar) and Gentamicin (R~Gent) from the second quarter to the fourth quarter.

Association Rule			*Confidence*			*χ^2 Analysis Result*
Left		**Right**	**Q2**	**Q3**	**Q4**	**Q2 to Q4**
Empty Set	=>	R~Ticar/Clav, R~Gent	4/209	9/292	13/263	0.954
			1.9%	3.1%	4.9%	

Figure 2 - Unconditional Associations

The finding is further supported by the discovery of a similar trend in the increase in significance, as shown in Figure 3, when the confidence of the conditional rule R~Gent => R~Ticar/Clav is being analyzed.

Association Rule			Confidence			χ^2 Analysis Result
Left		**Right**	**Q2**	**Q3**	**Q4**	**Q2 to Q4**
R~Gent	=>	R~Ticar/Clav	4/23	9/29	13/33	0.955
			17.4%	31%	39.4%	

Figure 3 - Conditional Associations

The discovery and the analysis of both the unconditional and the conditional association rules provides a warning that *Pseudomonas aeruginosa has* potentially developed resistance to the Ticarcillin/Clavulanic acid and Gentamicin combination. Hence, a review of the use of the antibiotic pair may be beneficial to the hospital.

To effectively utilize the alerts generated by Hawkeye in domains such as epidemiologic surveillance, a community of experts is needed to work collaboratively to provide authoritative interpretations of the alerts. An example of such a community of experts in local hospitals is an infection control group. The collaborative extension of the individual user is discussed first in the local hospital infection control context and then later in a geographically dispersed context.

4. Local Site Collaborative Interpretation

Timely interpretation of data is important to allow hospitals to make proper decisions and take actions to prevent and control potential outbreaks of disease or antimicrobial resistance. Preventive actions lead to quality patient care and more effective use of resources. Expertise in multiple fields is often necessary in order to perform the data interpretation in a timely manner to affect decision making and outcomes. The required expertise is usually distributed among multiple individuals. Alerts that are not considered important by one expert may be significant when viewed by another due to the diversity of knowledge and experience. Collaboration among the experts during interpretation would lead to meaningful screening of the alerts and would reduce potential misses of important ones. The result of the collaborative effort is an authoritative interpretation of the alerts that can effect decisions and policies. In the local hospital where Hawkeye is currently applied, the members of the hospital Infection Control Committee (ICC) are the experts who perform this interpretation.

The Infection Control Committee is chartered to improve the quality of patient care. One of the committee's responsibilities is the surveillance of infection activities in the hospital. The Clinical Microbiology Laboratory Information

System is an essential source of data for members of this committee to monitor infection occurrences and to identify antimicrobial susceptibilities. However, with existing database systems, users can only query the database according to an existing hypothesis, e.g. frequency of antibiotic resistance in a particular microbe. Then the users would review the results, which likely are long listings of data, to seek the information that they are interested in. The laborious nature of the process, coupled with the voluminous data, keeps many from tapping the potential of the data in the Laboratory Information System. Therefore, there is great need for tools to extract and present useful data in a compact form so that the Infection Control Committee can make informed, timely decisions to improve the quality of care.

Hawkeye's strength is in identifying patterns that could be of significance but not in the existing hypothesis set that the Committee is monitoring (i.e. patterns not known and hence not being tracked). Such capability offers the hospital the greatest potential to be proactive in preventing infections and hence improving care. As discussed earlier, to interpret the newly discovered patterns in a timely manner would require collaboration of the experts on the Infection Control Committee. Since these experts in general are scattered around the hospital and have tight schedules, a computer supported cooperative environment could be used to facilitate a timely collaborative interpretation process. Figure 4 depicts the computer-supported environment for collaborative interpretation of newly discovered epidemiologic knowledge at a local hospital site.

This diagram shows how data collected from various sources is mined using the Hawkeye system to produce new patterns, which are then presented to the users as alerts. The Computer Supported Cooperative Environment interfaces with the Hawkeye system to provide an environment to a group of experts to review and interpret these alerts. The environment will facilitate expert consensus about which alerts are useful and their interpretation. The environment will assume the moderator's role in the process. The resulting interpretations will then be archived in a Hospital Knowledge Repository. The content of the repository will in turn be used as input to Hawkeye to improve the selection capability of the discovery process.

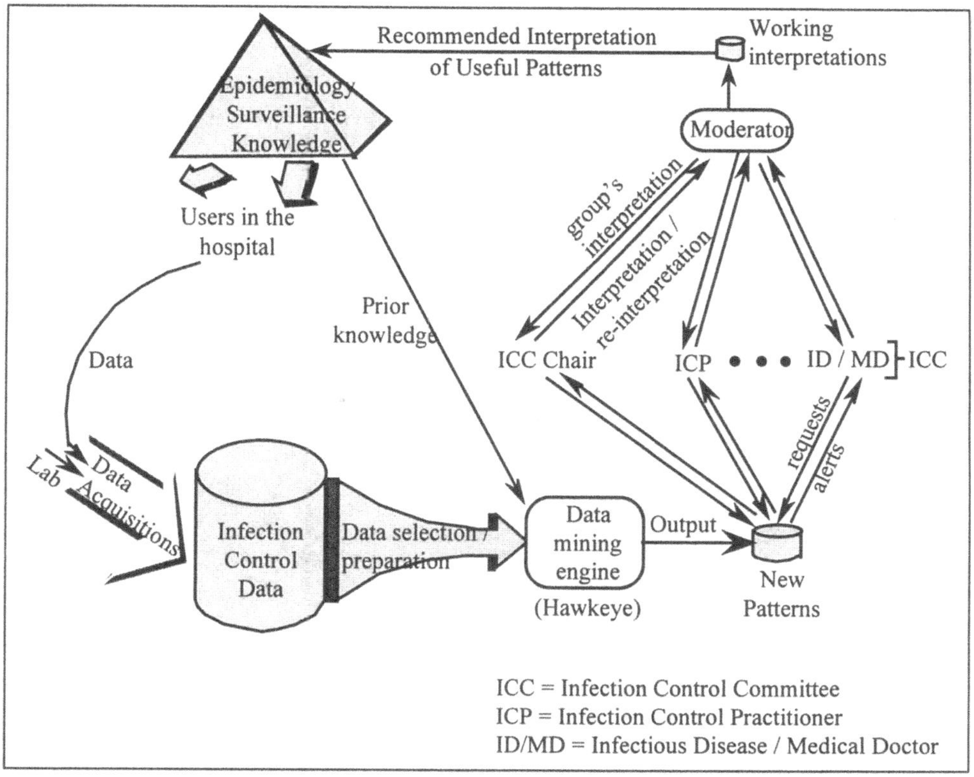

Figure 4 - Computer Supported Cooperative Environment for Collaborative Interpretation at a Local Hospital Site

The development of computer support for collaborative interpretation will take into account the following factors:

1. elicitation of each member's opinion
2. sharing the opinions of each member with the group,
3. forming an aggregate interpretation with consensus.
4. tracking difference in opinion so that contrary view points are not lost

Group decision support techniques such as Delphi study [5] are being explored as tools for facilitating the collaborative interpretation process with the computer as the moderator.

5. Toward Global Collaborative Interpretation

With rapid transportation and a mobile society, diseases can now spread rapidly from one area to other places of the world. Medical professionals in a newly affected area may not be knowledgeable in diagnosing and treating these diseases.

Therefore, collaboration of experts among hospitals within a region, a nation or from around the world could facilitate timely interpretation of important patterns and hence early diagnosis of an emerging disease that is formerly unknown to that area/site. In addition, experience in handling and preventing such diseases can also be shared through collaboration. The conceptual model proposed in Figure 5 is an extension of Figure 4 and enables a hospital to collaborate with other sites.

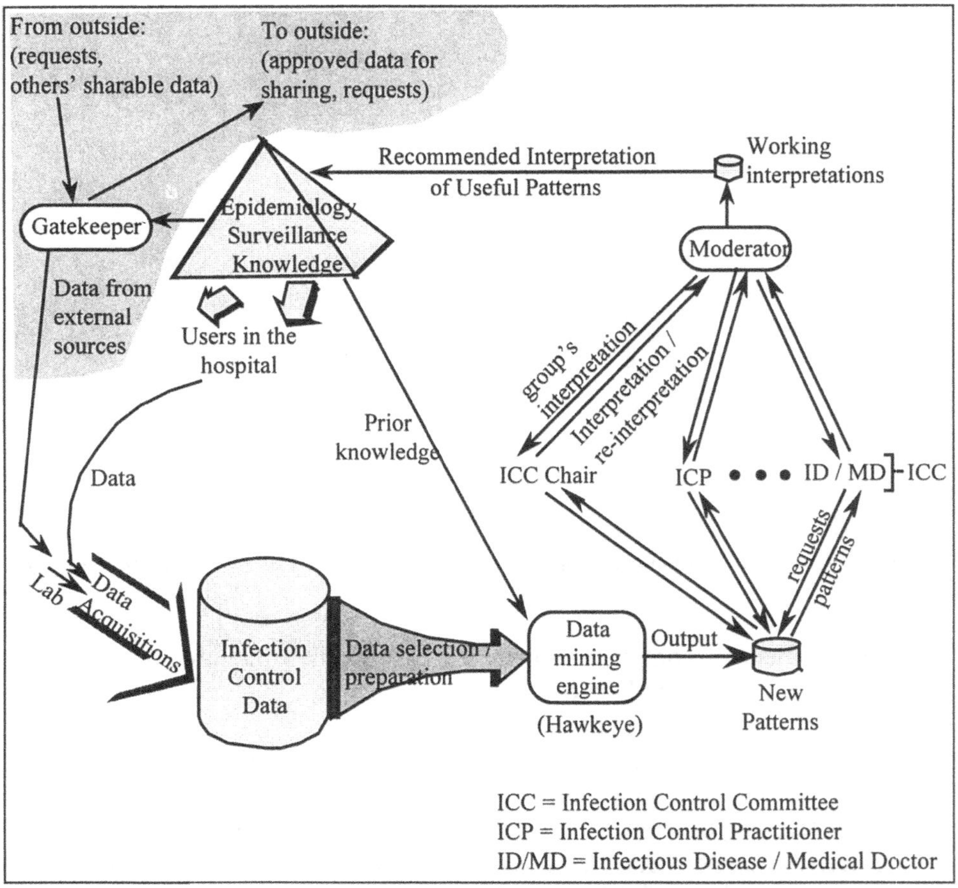

Figure 5 - A Local Site Model for Global Collaboration

Figure 6 provides a global view of hospitals leveraging on the Internet and the World Wide Web (WWW) technology. The Internet and the WWW are well established as channels for global information sharing and communication. It is beneficial to implement the Collaborative Interpretation model using user interfaces and communication protocols that are already established. For primitive sites that do not have the resources (e.g. equipment) to access the

WWW, simple email can be used to communicate to one of the regional gatekeepers.

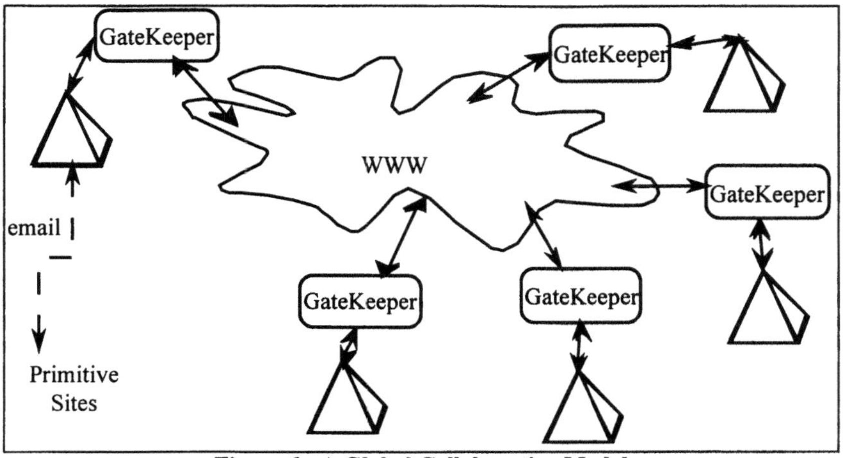

Figure 6 - A Global Collaborative Model

In the global model, security and privacy is of major concern. As shown in Figure 6, the communication from a site to the WWW is via a GateKeeper that knows what site knowledge is authorized to share externally. The authorization power is in the hands of the Infection Control Committee and the appropriate hospital authority. The GateKeeper will also be trained to filter the incoming information that is of interest to the hospital and add it to the regular data acquisition channel. Software agent technology could potentially be used to implement the function of the GateKeeper.

6. Use of Software Agents Technology

Intelligent software agent technology [6] is being explored for incorporation into various aspects of the aforementioned models. We summarize the potential applications as follows:

6.1 Functional Agents

Purpose:

- provide a functional profile, serve as an assistant to the user who is assigned to the function. In the models, the users are the collaborating experts such as the members of the Infection Control Committee
- help new comers to get into their role faster

- provide continuity during personnel turn over. This is achieved by keeping a current profile of the job function and growing an Organizational Knowledge Repository
- allow personalization to suit individual presentation preferences and working style

6.2 Moderator agent

Purpose:

- serve as the group moderator
- coordinate the group to come up with an aggregate recommendation on the interpretation of the alerts being reviewed (e.g. useful and promote to the Knowledge Repository but for internal use only; or the alert is irrelevant)
- maintain the Working Interpretation repository

6.3 Gatekeeper agent

Purpose:

- communicate with external sites - to send and to receive requests for information
- determine if external requests can be satisfied - according to whether the information related to the requests is available and whether the information is classified as for internal use only
- feed data collected from outside to the data acquisition channel
- collaborate with the Moderator agent to determine if the external information just received is related to any of the alerts under discussion. If appropriate, feeds the external information to the moderator immediately to support the current interpretation process, as well as sending the data to the acquisition channel

7. Conclusions

A framework of models has been presented for the application of advanced computer technologies such as knowledge discovery, group decision support systems and intelligent collaborative agents to the problems of local, national and global epidemiologic surveillance. Local site models are assumed to have a knowledge discovery tool such as Hawkeye with the capability for rapid, automated detection of unanticipated patterns and trends of interest in epidemiologic data. This framework meets important criteria recommended by ProMED such as a strongly horizontal decentralized structure with an emphasis on the development of powerful collaborative tools for supporting cooperation among sites.

8. References

1. Neu HC et. al. (1992) "*Antibiotic Resistance: Epidemiology and Therapeutics*", Diagnostic Microbiology and Infectious Disease (15) 53s-60s.

2. Brossette, SE; Moser, SA; Sprague, AP; Hardin, MJ, Jones, WT (1996) "*A Knowledge Discovery and Data Mining Surveillance System Applied to Infection Control*", UAB Department of Computer and Information Sciences Report CS-TR-96-11, Dec. 1996.

3. "*Global Monitoring of Emerging Diseases: Design for a Demonstration Program*", ProMED Paper, Federation of American Scientists.

4. Fayyad, UM; Piatetsky-Shapiro, G; Smyth, P (1996), "*From Data Mining to Knowledge Discovery: An Overview*", in Advances in Knowledge Discovery and Data Mining, ed. Fayyad, UM; Piatetsky-Shapiro, G; Smyth, P; and Uthurusamy, R. , MIT Press: Cambridge, Massachusetts.

5. Martinio, Joseph (1993), "*Delphi*", Chapter 2, in Technological Forecasting for Decision Making, Third Edition, pp. 15-35, McGraw-Hill, Inc., New York, New York.

6. Baecker, RM, Grudin, J, Buxton, W, Greenberg, S (1995), "*From Customizable Systems to Intelligent Agents*", in Human-Computer Interaction: Toward the Year 2000, pp. 783-793, Morgan Kaufmann Publishers, Inc., San Francisco, California.

7. Brossette, SE, Sprague, AP, Hardin, M, Waites, KB, Jones, WT, Moser, SA, (1997), "*Association Rules, Data Mining in Hospital Infection Control and Public Health Surveillance*", UAB Department of Computer and Information Sciences Report CS-TR-97, Sept. 1997.

An Early Warning System for Detection and Prediction of Outbreaks of Epidemics *

M. Bull G. Kundt L. Gierl
University of Rostock
Department for Medical Informatics and Biometry
Rembrandstr. 16/17
D-18055 Rostock
Germany
{mathias.bull|guenther.kundt|lothar.gierl}@medizin.uni-rostock.de

Abstract

In this paper, we present the methodology and the architecture of an early warning system for (1) detection of health risks, (2) prediction of the temporal and spatial spread of epidemics and (3) estimation of the consequences of an epidemic w.r.t. the personnel load and costs of the public health service. To cope this three task, we combine methods from statistics and artificial intelligence.

1 Introduction

In 1928, A. Fleming discovered the sensivity of pathogens to penicillin. Since that time, many antibiotics and vaccines have been developed and applied successfully. Nevertheless, the danger of epidemics has not been banished. Drug-resistant pathogens, e.g. vancomycin-resistant enterococci, the outbreak of the Ebola-virus in Africa, the world wide AIDS epidemic, or the relationship of BSE and Creutzfeld-Jacob-Disease are only a small fraction of new and re-emerging communicable diseases.

Physians, hospitals, and health authorities do not have access to an appropriate instrument which provides information needed for a timely decision making from the huge body of medical data avaiable. In this paper, we present our approach to the problem of health surveillance - an early warning system. This is a part of the project TeCoMed, a Tele-Consultation-System for acute regional health risks, which we establish in the state Mecklenburg-Vorpommern (north-east part of Germany).

*supported by the DFN-Verein (German society for the national research net) and the AOK Mecklenburg-Vorpommern (General health insurance company)

2 Case-Based Reasoning in Epidemiology

The public health surveillance is a complex problem of multiparametric time courses of diseases, pathogens, resistances, health services etc. in a geographical region. Often, the epidemiological relationships are unknown, undiscovered, or only weak examined. Consequently, the modelling of a correspondence Φ between a given multiparametric feature x and a required solution y is very complicated or impossible.

$$x = (x_1, x_2, \ldots, x_m) \xrightarrow{\Phi} y = (y_1, y_2, \ldots, y_n)$$

Analogical reasoning is an approach to overcome this circumstance based on the philosophy: similar problem instances have similar solutions. Here, the modelling of Φ is reduced to define a suitable similarity measure and to create an adaptation mechanism.

$$\begin{array}{lcl}
\textit{known problem–solution–pair} & & \\
x' = (x'_1, x'_2, \ldots, x'_m) & \longmapsto & y' = (y'_1, y'_2, \ldots, y'_n) \\
\uparrow \textit{similarity} & & \downarrow \textit{adaptation} \\
x = (x_1, x_2, \ldots, x_m) & & \tilde{y} = (\tilde{y}_1, \tilde{y}_2, \ldots, \tilde{y}_n) \\
\textit{current problem instance} & &
\end{array}$$

Case-based reasoning (CBR) is an intelligent combination for solving problems by analogical reasoning and for learning new problem-solution-pairs by experience. The CBR-cycle (see [1]) consists of the following four stages: *retrieve* the most similar case or cases; *reuse* the retrieved case or cases to solve the problem by analogical reasoning; *revise* the proposed solution; *retain* the parts of experience likely to be useful for future problem solving.

There are successful applications which use CBR-techniques for monitoring complex processes in time and space in other domains (cf.[8], [9], [10], [11], [16]).

In geographical epidemiology, G. F. Pyle [14] compared the diffusion paths of influenza epidemics and found out that there exist similar epidemic waves according to the time-spatial dynamic. In [12] R. J. Marshall discussed various methods for analysis of geographical distribution of diseases and mentioned that there are observable spatio-temporal patterns of the spread of diseases which are complicate to describe by means of statistics (especially see [12], section 5).

Following the argumentation given above, we combine methods of statistics and artifical intelligence (AI), namely case-based reasoning, knowledge discovery and data mining (KDD) to cope with the problem of health surveillance.

3 System Architecture of TeCoMed

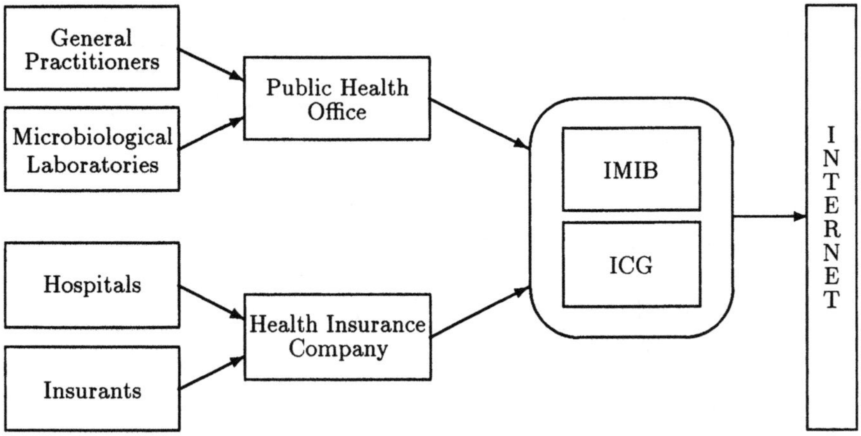

Figure 1. Information flow of TeCoMed

The crucial point with the analysis of the time-spatial pattern of diseases and pathogens is the reliability, the validity, and the timeliness of the data.

A main principle of our approach is that we use various data sources of medical data which are already available as formatted data on machine-readable mediums. In Germany, the hospitals have to collect patient records on machine-readable mediums (§301 SGB V, §301-agreement on the transmission of medical data) and have to send these records to the health insurance companies. Moreover, both patient and general practitioner have to send the medical certificate within three days to the health insurance companies. We daily recieve all these data as anonymous patient records (ca. 2200 records about medical certificates and ca. 700 records about deliveries to hospitals per day) via net from the computing center of general health insurance company "AOK Mecklenburg-Vorpommern" which insures about 40% the population. These records contain the following items: decade of the birth year (e.g. sixtieth), sex, encoded profession, postcode of the domicile (5 digits), date of fall ill, emergency, convalescence, death, and the diagnosis given by ICD-code. The ICD-code (International Statistical Classification of Diseases and Related Health Problems) is a hierarchy of disease clusters (e.g. ICD9: 460-519 Diseases of the Respitory System, 487 Influenza). Furthermore, the public health office "Landeshygieneinstitut Mecklenburg-Vorpommern" weekly sends information about notifiable diseases and about pathogens identified by the microbiological laboratories.

The gathered data of hospitals, health insurances and microbiological laboratories are transmitted to servers at the Institute for Medical Informatics and Biometry (IMIB) and at the Institute of Computer Graphics (ICG) of the University of Rostock (see Figure 1). Then, the procedures of data abstraction, health risk detection and prediction start which we describe in following sections. The results are integrated in the knowledge base. The visualization tools MedVis (see [17]) generate graphical representations of the data which shows the detected health risks and the spread of communicable diseases as well as the temporal and spatial dependencies and correlations of the data.

In the next future, the functionality of the visualization system will be used via the net without having the actual visualization system implemented on the user's terminal or network.

4 Abstraction of Health Data

In this section we describe the abstraction mechanisms from the raw data to spatio-temporal knowledge structures, called scenario sequences.

4.1 From Data to Scenario

By a *scenario* we understand a concept which describes the public health situation in the considered region during a time period. From the raw data described in section 3 shortly, we construct a *scenario* as follows (see Figure 2 as illustration).

Spatial data abstraction by accumulation
Firstly, we choose a reasonable period τ (by organisational reasons $\tau=$ one week) and divide the time scale in equidistant time steps. The geographical region, state Mecklenburg-Vorpommern, is divived in districts and further in postcode units. These dissections of the region form a location hierarchy: state - districts - postcode units. Then, for each postcode unit we accumulate the new cases with respect to the disease, to the birth decade, and to the sex.

Spatial data abstraction by identifying of distribution patterns
In a second step, we explore disease distribution pattern by case-based and statistical methods.
Although we have a large and rich data sources, we have to notice that these data can not be considered as a representative sample within the meaning of statistics. The social structure as well as the age distribution of the insurants and of the whole population do not correspond. Consequently, parametric statistical methods are not applicable. Therefore we combine case-based

techniques and permutation tests. Permutation tests (see [7], [3]) are special distribution-free statistical tests which require only one or two relatively weak assumptions about the distribution of the variables, e.g. the underlying distribution is symmetric.
The G-statistic is based on permutation tests which measures the concentration or the lack of concentration of the sum of considered values in the region. This statistic is a proportion of the sum of all values that are within a certain neighbourhood of a location to the total sum of the region. Whether this G-statistic value for a location is statistically significant depends on the statistic's distribution. The null hypothese is that the set of all values within a certain neighbourhood of a location is a random sample drawn without replacement from the set of all values in the region. We refer the reader to the interesting paper [6] by Getis and Ord.

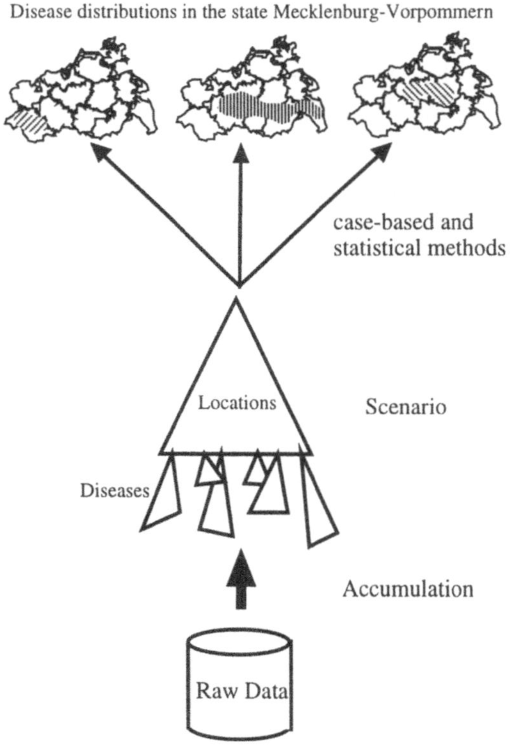

Figure 2.

4.2 From Scenarios to Scenario Sequence

Obviously, by a *scenario sequence* we mean a sequence of scenarios. We also denote a scenario sequence by the concatenation of its elements, e.g. $\sigma =$

$s_0 s_1 \ldots s_n$ where every s_i, $i \in \{0, 1, \ldots, n\}$, is a scenario. So we are able to describe the course of any epidemic by a scenario sequence since it keeps all information on the public health situation during a epidemic period.

Spatio-temporal data abstraction by scenario concatenation
With respect to the a considered disease or cluster of diseases, we concate consecutive scenarios, s_i, to a scenario sequence σ. Here, we start with that scenario which contains an observable pattern (according to considered disease) and is the first one with respect to time.

Spatio-temporal data abstraction by spread summary
We summarize the spread of an epidemic over the time in a so called *spread summary*, that is a list containing the locations with epidemic level in their temporal order. This is illustrated in Figure 3. All detected epidemic scenario sequences we have organized in a case base Σ, the geomedical knowledge base.

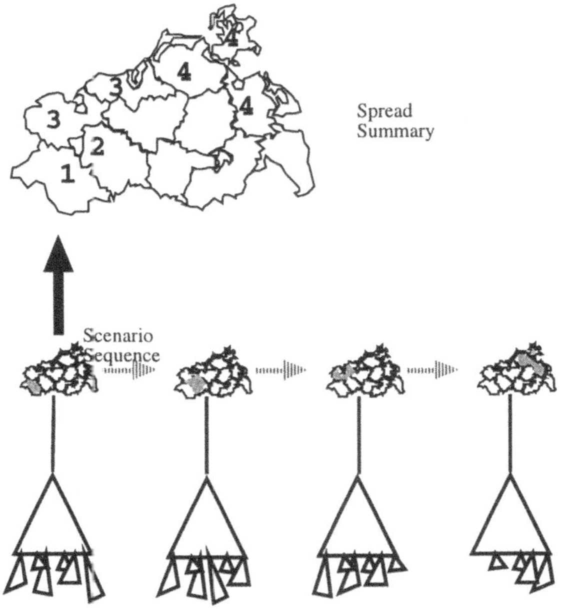

Figure 3.

5 Case-Based Risk Detection and Prediction

In this section we sketch the methodology of health risk detection and prediction by means of case-based reasoning. For that, we define inductively a similarity metric *sim* on the scenario sequences, i.e. at first we define similarity metrics on all elementary sets and then similarity metrics on sets of structured objects by using the metrics of its components.

By a *health risk* we understand an observable change of the public health situation which may cause an epidemic. So we have the following main problems:

Health Risk Detection
Given: σ current scenario sequence
Question: Does there exists a health risk ?

Health Risk Prediction
Given: σ current scenario sequence
Question: How does the successor scenario of σ looks like ?

Estimation the Consequences for Health Care Resources
Given: σ current scenario sequence
Question: What are the consequences for health care workers, pharmaceutical industry and over all costs ?

In the following sections we denote the case base by Σ which contains some scenario sequences describing the course of past epidemics. Let $\sigma = s_0 s_1 \dots s_m$ be the current scenario sequence.

5.1 Health Risk Detection

The discovery of a health risk is based on the question "Has something changed in a suspicious manner?" Here, conspicious changes in the health situation of the inhabitants of a region could be an accumulation of an illness of undetermined origin within a certain time interval, a suspect accumulation of resistances to antibiotics in a certain area, or a sickness which occurs frequently within a period of time or a certain area. Remark that an accumulation can also occur with a very low incidence rate of an event. Therefore, it is nescessary to determine each minor change of the health situation. We want to detect a health risk as soon as possible to warn whereas in epidemiological studies the aim is to prove a causal relationship of diseases and expositions, this means in our problem a pathogen. However, a frequent false-positive reaction of the system must be avoided (artifacts). It must be possible to set the sensitivity and specificity of the system.

Discover conspicuous changes in the current scenario sequence
By using methods of KDD (cf. [5], [13]) the current scenario sequence σ is explored. Especially, for early detection of outbreaks we apply also methods

described in [4]. Moreover, existing techniques of Scientific Visualisation can be applied to show temporal and spatial dependencies and correlations of the data contained in the current scenario sequence σ.

Retrieve the scenario sequences with similar onsets and epidemic courses
The system looks for a scenario sequence σ' which starts with a subsequence similar to the current scenario sequence σ and terminates in a dangerous scenario. (Remark that a pattern of a conspicuous changing must not indicate an outbreak.)

Reuse the retrieved scenario sequences to warn
If there is such a scenario sequence σ' then the system has to warn or alarm of an epidemic depending on severity degree of the detected health risk.

epidemic scenario $\swarrow$

retrieved scenario sequence from the case base Σ $\longrightarrow \sigma' = s'_0 s'_1 \ldots s'_m s'_{m+1} \ldots s'_{m+k}$

$\Downarrow$ *warning*

current scenario sequence $\longrightarrow \sigma = \underbrace{s_0 s_1 \ldots s_m}_{\text{similar subsequences}} \boxed{!}$

The severity degree of a health risk is characterized by the incidence, population density of a location (e.g. whether the outbreak occurs in a city or village), quality of a prevention (e.g whether there is a vaccine), quality of a therapy (e.g. how successful will be a medical treatment) course of the illness.

Revise the warning
The quality of the discovery of health risks is evaluated by monitoring of the succeeding scenarios.

Retain the useful scenario sequences
If a scenario sequence which led to an epidemic and has not been discovered by the system then this sequences will be stored in the case base.

5.2 Health Risk Prediction

For prediction of a regional health risk, i.e. to forecast a future scenario, we use the following steps.

Retrieve the most similar scenario sequences
The system retrieves all scenario sequences $\sigma' s'_{m+1} = s'_0 s'_1 \ldots s'_m s'_{m+1}$ from the case base Σ which are similar to the current one, σ, i.e. for all $\sigma'' s'' \in \Sigma$ it holds $sim(\sigma, \sigma'') \leq sim(\sigma, \sigma')$. Here, all epidemic scenario sequences are indexed by

the diseases, by the outbreak locations and by the spread summary of the epidemics as described in section 4.1 and 4.2.

Reuse the retrieved scenario sequences to forecast a scenario
By using the retrieved sequences $\sigma' s'_{m+1}$ the system adapts the scenario s'_{m+1} to a forecast scenario $\tilde{s}$.

$$
\begin{array}{lcccl}
\text{similar scenario sequence} & & & & \\
\quad\text{from the case base } \Sigma & \longrightarrow & \sigma' s'_{m+1} & = & s'_0 s'_1 \ldots s'_m s'_{m+1} \\
 & & & & \qquad\quad \Downarrow \textit{adaptation} \\
\text{current scenario sequence} & \longrightarrow & \sigma & = & s_0 s_1 \ldots s_m \boxed{?} \\
 & & & & \qquad\qquad \nwarrow \\
 & & & & \text{predicted scenario } \tilde{s}
\end{array}
$$

Here, we use background knowledge about demographic structures, several statistical models of epidemics (see [15]) and the defined threshold values of epidemical levels which are based on the epidemiological studies. The resulting data will be stored in a case based fashion and will be available for graphical presentation on the user interface.

Revise the forecast scenario
After a time period τ we know the scenario s_{m+1}. In comparing the predicted scenario $\tilde{s}$ with the scenario s_{m+1} the system evaluates the prediction mechanism.

Retain the useful scenario sequences
If the difference of these scenarios is too large then the system has to learn this new scenario sequence σs_{m+1} or the inference mechanism must be changed. In the first case, the current scenario sequence is integrated in the case base. In the latter, an expert (e.g.epidemiologist, biostatician) has to execute a working cycle with the system. This work cycle contains the modifying of the threshold values and of the metrics, the visualizing of the multiparametric data of the current scenario sequence and those in the case base and the forecasting of a scenario (as a simulation step). This cycle is executed until the difference of the current scenario and the predicted senario is within a tolerance range.

5.3 Estimation the Consequences for Health Care Resources

Based on the discovered risk and on the forecast scenario, monitrary and medical consequences will be estimated. Furthermore, the demands on intensive care beds, the required nursing care or amount of vaccines can be concluded. Length of stay, possible surgery, letal cases and further consequences can also be deduced. From consumables, departmental care rates etc. the expected costs for a case induced by the epidemic can be estimated.

To illustrate the economic impact of an epidemic, we notice that the influenza epidemic 1995/1996 caused 900 million DM only for medical treatment in Germany.

6 Conclusion

Presently, there are some health surveillance networks as singular solutions for observing special chosen or notifiable communicable diseases, which can be classified in three groups: Sentinel-General-Practitioner-Networks, Sentinel-Microbiological-Laboratory-Networks and E-Mailing-Lists. In all these health surveillance systems the information gathering, the information flow, the information processing and the publication of the monitoring results are separate from the offical report channels. Consequently, these systems do not improve the information management within the networks of the offical report channels.

In our approach, the information gathering is based on a well-designed network using the data which already exist on machine-readable files. For the information processing, we combine and implement methods form several fields, namely statistics, artificial intelligence, and scientific visualisation. For a rapid publication of the monitoring results we utilize the facilities of the INTERNET.

References

[1] Aamodt, A. ; Plaza, P. : *Case-Based Reasoning: Foundational Issues, Methodological Variations, and System Approaches.* AI Communications 7(1) 1994, 39-59

[2] Bailey, N.J.T.: *The Mathematical Theory of Epidemics.* High Wycombe, Charles Griffin, 1957

[3] Edgington, E.S.: *Randomization Tests.* Marcel Dekker, Inc., NewYork 1987

[4] Farrington, C.P.; Andrews, N.J.; Beale, A.D.; Catchpole, M.A.: *A Statistical Algorithm for the Early Detection of Outbreak of Infectious Diseases.* J.R. Statist. Soc. A (1996) 159, Part 3, 547-563

[5] Fayyad, U. M. ; Piatetsky-Shapiro, G.; Smyth,P. ; Uthurusamy, R.: *Advances in Knowledge Discovery and Data Mining.* MIT Press, 1996

[6] Getis,A.; Ord, J.K.: *The Analysis of Spatial Association by Use of Distance Statistics.* Geographical Analysis, Vol 24, N0.3 (July 1992), 189-205

[7] Good, Ph.: *Permutation Tests.* Springer-Verlag, Berlin 1986

[8] Jones, E.K. ; Roydhouse, A. : *Iterative Design of Case Retrieval Systems.* Victoria University of Welligton, New Zealand, CS-TR-94/6, see also: Proc.of the AAAI-94, Workshop on CBR, Seattle, Washington, 1994

[9] Kitano, H.; Shimazu, H.; Shibata, A.: *A Methodology for Building Large-Scale Case-Based Systems.* in: Proc. of the AAAI-93, Washington, 1993

[10] Lee, J.K.; Oh, S.B.; Shin, J.C.: *Unik-FCST: Knowledge-assisted adjustment of statistical forecast.* Expert Systems with Applications, Vol. 1, 39-49

[11] Lekkas, G.P.; Arouris, N.M.; Viras, L.L.: *Case-Based Reasoning in Environmental Monitoring Applications.* Applied Artificial Intelligence, Vol. 8, 1994, 349-376

[12] Marshall, R.J.: *A Review of the Statistical Analysis of Spatial Patterns of Disease.* J. R. Statist. Soc. A (1991) 154, Part 3, 421-441

[13] Piatetsky-Shapiro, G.; Frawley, W.J. (Eds.): *Knowledge Discovery in Databases.* AAAI Press, Menlo Park, 1991

[14] Pyle, G.F.: *The Diffusion of Influenza.* Rowman & Littlefield, Totowa NJ 1986

[15] Rothman, K.J.: *Modern Epidemiology.* Little Brown, Boston, Toronto 1986

[16] Schmidt, R.; Heindl, B.; Pollwein, B.; Gierl, L.: *Abstraction of Data and Time for Prognoses of the Kidney Function in a Case-Based Reasoning System.* in J. Brender, J.P. Christensen et al.(eds.): Proc. of Medical Informatics Europe '96, IOS Press, Technology and Informatics, Vol. 34, Part A, 570-574

[17] Schumann, H.; Lopez de Chavez, N.; Graw, K.U.: *Visual Analysis as an efficient Tool for the Exploration of Human Health Data.* in this volume

Visual Analysis as an efficient Tool for the Exploration of Human Health Data

H. Schumann[a] N. López de Chávez[a] K.-U. Graw[b]

[a]University of Rostock, Institute of Computer Graphics 18051 Rostock, Germany
e-mail: schumann,lopez @informatik.uni-rostock.de

[b]Fraunhofer Institute for Computer Graphics Rostock, J.-Jungius-Str. 9, 18059 Rostock
e-mail: kugraw@egd.igd.fhg.de

Abstract

To ensure a high level of human health protection the huge amount of data gathered day by day by the physicians of the health service have to be analyzed in an efficient way. The current statistical and numerical methods allow only a limited analysis of these data. In this paper we discuss the visualization of huge data sets of multiparameter health data with geographical and temporal dependence as an efficient tool for the analysis of human health data and present some new solutions. Furthermore, we describe the visualization system MedVis which has been developed for the visualization of health data of the German state Mecklenburg-Vorpommern as part of a tele-consultation-system for the monitoring of acute health risk.

1 Introduction

Day by day a huge amount of medical data from patients which describe their diseases and their treatment is collected by the physicians of the health service. In order to ensure a high level of human health protection by taking effective precautions in time the analysis of these data becomes more and more important. Especially preventive actions against communicable diseases may prevent the outbreak and spreading of epidemics.
In Germany the current information flow is too slow and the analysis of these data consists mainly of weekly, monthly and yearly reports with simple tables of the data published by the regional public health departments and the Robert-Koch-Institute. Especially for the prevention of epidemics a faster and better monitoring and analysis of the data for an early warning system is necessary.
Therefore we have to use better means of communication (WWW) and more effective analysis methods, e.g. Scientific Visualization. Especially graphical representations allow an intuitive analysis of the data for different usergroups and offer an alternative way to the tedious and time consuming extraction of information from statistical tables.

2 Problem Description

All notifiable diseases have to be recorded by the physicians and are reported daily or weekly to the Public Health Department of Mecklenburg-Vorpommern. These data form the data set

$$\mathcal{MD} = \left\{ \left\{ \begin{matrix} k_{11} & \dots & k_{1k} \\ \vdots & \dots & \vdots \\ k_{m1} & \dots & k_{mk} \end{matrix} \right\}_{t_j}, \; t_j \in [t_a, t_e] \right\}$$

where k_{ij} denotes the frequency of the disease K_i in the geographical area o_j, e.g. district or district town, and $[t_a, t_e]$ denotes the time interval.

Another data source are the patients data recorded by the health insurance companies according to the Health Structure Law §301. These data describe every medical case in detail and contain specific information like age and sex of the patient or length of illness and the kind of treatment. These data form the data set

$$\mathcal{CD} = \left\{ f_{o_i, t_j}^{K_l, m_1, \dots, m_k}, \; o_i \in \mathcal{O}, \; t_j \in [t_a, t_e], \; K_l \in \mathcal{K} \right\}$$

where $\mathcal{O}$ denotes the set of geographical areas, $\mathcal{K}$ the set of diseases and $m_1, \dots, m_k$ the parameters describing the disease and the patient. The data set $\mathcal{MD}$ can be easily extracted from $\mathcal{CD}$ and we have additional information for a more detailed analysis.

These data have to be analyzed within their spatial context. Therefore we need additional data which contain further semantic and demographic information about the geographical areas. Such necessary semantic information are e.g. Granularity structur, Neighbouring geographical areas. Furthermore we need demographic information about the geographical areas such as size of geographical areas, density of population, age structure.

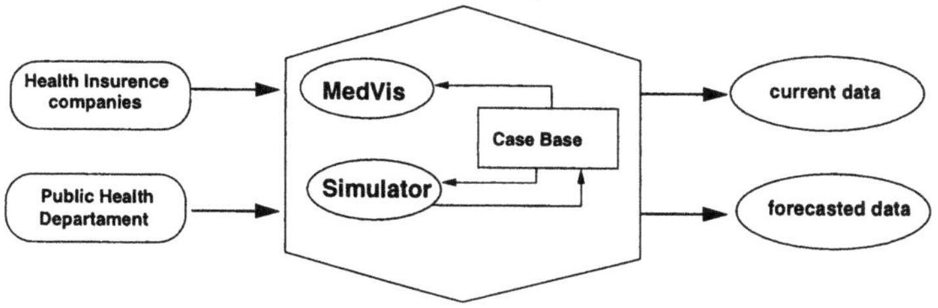

Figure 1: The Tele Consultation System

The visualization system MedVis is part of a tele-consultation-system which is developed in cooperation with the Institute of Medical Informatics and Biometrics in the project 'TeCoMed - Tele-Consultation-System for acute health risks in Mecklenburg-Vorpommern' sponsored by the DFN (see Fig.1). The task of the tele-consultation-system is to provide physicians and future health risks in a geographical region (see [Gierl96]). The input data $\mathcal{CD}$ for this system are updated weekley by the Health Insurance Companies and are converted into cases, i.e. sequences of the time series, which are classified into prototypes of cases. These data form the case base. The simulator searches for the nearest case to an current case in the case base by using the Case-Based-Reasoning (CBR) approach (see [Gierl97]) and tries to forecast the further course of the current case. The task of MedVis is the visualization of both the health data of Mecklenburg-Vorpommern and the current and forecasted cases of the case base and furtheremore its is realization via WWW.

3 The Role of Scientific Visualization

In the last few years the visualization of data has become a commonly accepted and widely used approach for the extraction of relevant information from huge abstract data sets. In many cases a better insight into complex processes and phenomena can be gained by means of graphical representations because the human brain allows the simultaneous perception of a large amount of optical information. Thus, a lot of information can be encoded in one single picture and therefore visual analysis offers an alternative way for the extraction of information to the more traditional approaches, like datamining or statistical analysis. Especially multiparameter data with temporal and spatial dependence can not be handled sufficiently by these traditional methods because of the large number of possible correlations. Scientific Visualization is concerned with the adequate transformation of abstract data sets into graphical representations which allow a better analysis, understanding and communication of data, models and concepts in science and engineering.

In our context the success of the visualization depends mainly on the fact whether the user can detect spatial and temporal relationships between the various parameters in the graphical representation and relate them to the underlying context. Therefore, all used visualization techniques have to be developed and implemented in close cooperation with the particular user groups. Since wrong conclusions caused by misinterpreted artifacts in the graphical representations are very dangerous in medical applications the graphical representations must not create falsification of the data and missing data must not be interpolated.

4 MedVis - A System for the visual Analysis of Human Health Data

4.1 Underlying Concepts

This section discusses the underlying concepts and methods which are necessary for an adequate visualization of the human health data.

4.1.1 Data Selection for visual Analysis of Health Data

The whole data contain too much information and can not be displayed in one picture. Furthermore, most users are only interested in a specific part of the data, e.g. a specific geographical area or time interval. Thus the amount of data to display has to be reduced by focusing. This can be done by either a *projection* or a *selection*.

A projection π reduces the amount of information by decreasing the number of parameters. The user chooses the parameters of the data set which are relevant for the visual analysis. Thus, he or she changes the structure of the datasets to display:

$$\pi : f_{o,t}^{K_l, m_1, \ldots, m_k} \longrightarrow f_{o,t}^{K_l, \tilde{m}_1, \ldots, \tilde{m}_{k_\pi}}, 0 \leq k_\pi < k,\ \tilde{m}_i \in \{m_1, \ldots, m_k\},\ i = 1, \ldots, k_\pi$$

A selection σ reduces the amount of information by decreasing the number of data sets to display.

$$\sigma : \qquad \mathcal{CD} \longrightarrow \mathcal{CD}_\sigma, \qquad |\mathcal{CD}_\sigma| < |\mathcal{CD}|$$

Focusing in the medical data can be done with regard to 3 different main criteria:

- Focusing in parameter space: projection of the parameters describing the disease and the patient, selection of diseases with regard to their parameter values
- Focusing in geographical space: choosing granularity level, selection of geographical areas
- Focusing in time: selection of times and time intervals

Whether to start focusing with regard to parameter space, geographical space or time depends mainly on the interpretation goals of the user, e.g. localization of a specific disease in a certain geographical area or time interval.

4.1.2 Visualization of Health Data

Many visualization techniques for the display of multiparameter data have been developed in the last years. In particular the *iconographic approach* is widely used, i.e. 'mapping data to an object - the icon - whose attributes are under the control of the various fields of the data record' [Grienstein95] . Moreover, icons can be positioned exactly and are therefore very suitable for the display of spatial relations. Other techniques are for instance *geometric techniques*,e.g. Hyperslice [WL93]and Parallel Coordinates [Inselberg94], *hierarchical techniques*,e.g. World-within-Worlds [FeBe90] and Treemap [Johnson93], among other.

Apart from the identification of extrema and frequencies for single parameter the identification of multivariate correlations is an important interpretation goal for the analysis of the health data and different visualization techniques have to be used for quantitative and qualitative analysis of correlations. The same icon can be used for the analysis of different kinds of correlations (see section 4.2), e.g. for the display of the frequency of one disease for different age groups for a fixed time and geographical area, several diseases for a fixed time and geographical area, one disease for different times and a fixed geographical area or one disease for a fixed time and different geographical areas.

4.1.3 Display of spatial Dependence

An adequate representation of the spatial frame of reference is necessary to assign important data features to their spatial context. Therefore the main concern is the display of the different spatial granularities. *Maps* are very suitable for the display of geographic spatial dependence because they have a low level of abstraction which allows an intuitive perception of spatial relations. Another possibility are *cartograms* (see [Dorling94]) which are 'maps in which a particular exaggeration is deliberately chosen', e.g. geographical units represented by Chernoff faces. However, when the number of parameters is large information can not always be displayed at a map. Therefore we distinguish in [Schumann96] between the direct or indirect display of spatial dependence.

Direct display of spatial dependence with maps and icons is more suitable for qualitative exploration of data with small numbers of data records and parameters whereas the indirect display is more suitable for data with larger numbers of data records and parameters. Furthermore, *indirect display* allows the use of other visualization techniques which are more suitable for quantitative analysis, e.g. Parallel Coordinates.

In our context we have to deal with different levels of granularity which require different visualization methods. On a rough granularity level there are usually small numbers of different geographical areas and the direct display with icons can be used efficiently. On a more detailed granularity level we can still use direct display if we zoom into the map and represent a smaller geographical region. However, the relation to the whole geographical region is lost. If we want to display a large geographical region on a more detailed granularity level we have to use indirect display instead.

On the other side the indirect display of spatial dependence is less intuitive. Therefore we have developed another approach. Instead of displaying the data corresponding to one geographical area with separate graphical representations we group areas according to their data and represent *groups of areas* with equal or similar data by one graphical representation, e.g. icon. Thus we reduce the number of graphical representations without loss of information and a direct display with icons could be possible. One icon represents the data of a group of several geographical areas and the areas which belong to a certain group can be indicated by a specific encoding, e.g. colour coding. Since the number of groups depends on the data the graphical representation should switch automatically between direct and indirect display depending on the number of groups.
How to display the spatial dependence and which granularity level to choose depends also on the spatial interpretation goals like :

-locating geographical areas with important data features
e.g. areas with a high incidence of diseases

-spatial course of an epidemic

outbreak area and spatial course of a communicable disease

The display of the spatial course of communicable diseases for representing the spreading of epidemics is difficult. The data are assigned to areas of different size and do not carry explicit direction information. A possible way offers the *maximum-icon* (see Fig. 2). The icon could be read as a clock representing a given time-intervall and the rand shows the time with the maximum of incidence of a disease to be anylized.

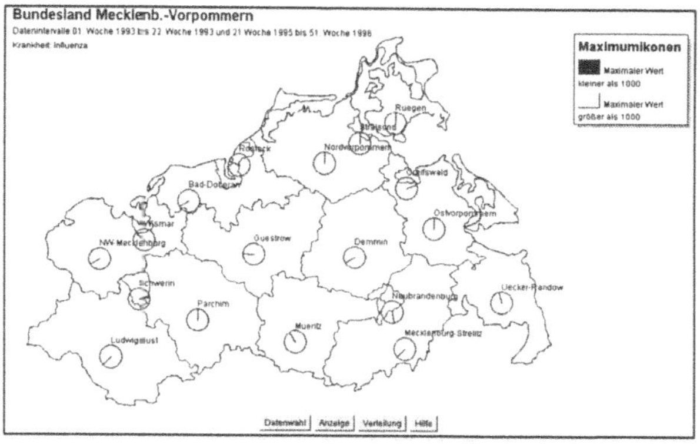

Figure 2: Maximum icon for the display of the spatial course of a disease

4.1.4 Display of temporal Dependence

Apart from the display of the spatial dependence an adequate display of the temporal dependence is necessary to detect significant changes and events. Thus we need visualization methods for identifying temporal changes by comparing static or dynamic representations for different times or time intervals

- Comparison of data for different times:
analysis of data for selected geographical areas and diseases at different times
- Comparison of temporal courses for time intervals:
visual analysis of temporal courses with regard to a specific disease for a geographical area at different time intervals or for different geographical areas for the same time interval.

The detection of a significant increase or decrease of a disease is an important goal and must be possible through visual analysis. Since a significant change for certain diseases can consist of only a minor change in the frequency it is difficult to display these small differences in an intuitive way especially when monitoring another disease at the same time where significant changes consist of big changes in the frequency. Therefore we want to use *events* defined on the time axis to display temporal courses in a more striking way. For this we want to use the concept of an *event dependent visualization.* Following [Liere96] an event dependent visualization is the display of dynamic processes where the graphical representation changes automatically when certain events occur. Events are emphasized explicitely in the graphical representation which allows an easier detection of significant changes. In figure 3 the two events, decreasing and increasing of incidences are distinguished by different icon types.

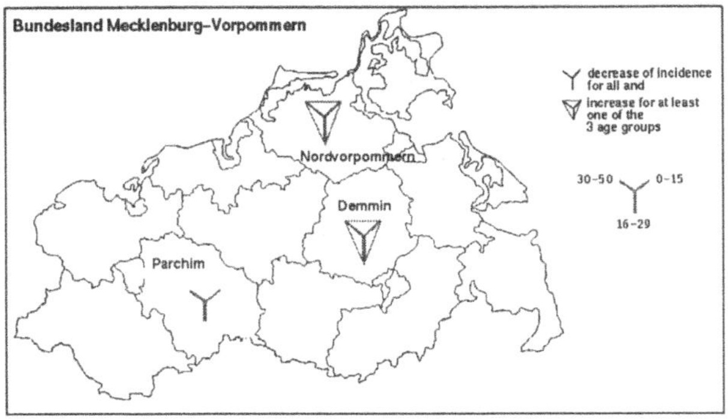

Figure 3: Event dependent visualization with 2 different icons

An event dependent visualization is especially suitable for our context where we have to locate data figures and events with are already known in the most cases. Furthermore, it allows the monitoring of several parameters on a rougher granularity level and thus the monitoring of a larger geographical area. It also reduces the cognitive demands significantly because only when an important event appears a graphical representation of the data is displayed instead of the more demanding visual analysis of a consistant graphical representation. However, we have to decide first which events and which shape of an event is going to change the graphical representation. Important events for our application have to be reconized by the simulator (see fig 1e.g.: fast and significant changes of a disease,exceeding of disease related threshold values, high incidence of a disease, occurrence of exceptional and dangerous diseases. The alteration of the visual representation caused by an event can consist of: different graphical attributes, e.g. different colour table, different icons for different events, different geometrical primitives

4.2 Realization

The visualizationsystem MedVis has been developed for the visual analysis of medical multiparameter data for the German state Mecklenburg-Vorpommern and contains implementations of some of the concepts introduced in section 4.1 especially for the visualization of huge data sets with spatial dependence in different granularities. Visualization techniques and focusing and interactive functionality have been combined to an efficient tool for the visual analysis and presentation of health data.

MedVis is based on the visualizationsystem IRIS-Explorer which allows the design of complex visualization applications from a set of modules. Furthermore, the functionality of IRIS-Explorer can be extended with own modules, e.g. we have developed modules for additional visualization techniques, focusing functionality and the input and visualization of maps. We are using the NetCDF-format (see [RDE93]) for data input, which allows also the storage of additional meta information about the data, e.g. maxima and minima of parameters, and preprocessed data, e.g. numerical correlation information. These information can be used for a more adequate graphical representation of the data.

4.2.1 Focusing in the Health Data

Since IRIS-Explorer is not linked to a data base we had to implement focusing functionality. The diseases and their parameters can be specified for visualization interactively from simple lists or with explicit input of the parameter name by the user. The same method is used for the specification of time steps and intervals. The user can choose between different time steps, e.g. day, week, month, quartal and year, and the data is automatically scaled according to the chosen time step. Geographical areas can also be selected from simple hierarchical lists, by explicit input or by picking in the graphical representation.

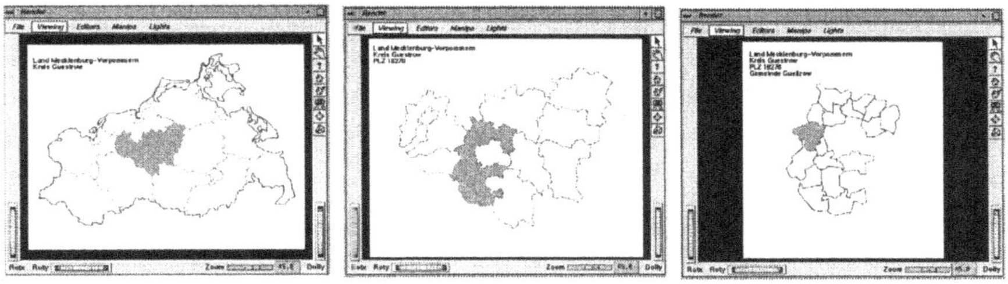

Figure 4: navigation through the 3 different granularity levels

The map of Mecklenburg-Vorpommern can be displayed in 3 different granularity levels, i.e. district and district towns, post code area and communities. The user can select the granularity level interactively and decide whether the whole region or only the selected areas are displayed. The contours of the geographical areas of different granularity levels are displayed in different colours to distinguish different granularity levels for a large region. On each granularity level each geographical area can be picked separately for navigation on the map and specification of the graphical representations to display (see Fig. 4). The size of the selected geographical region, the granularity level and the number of specified areas influence the way how the spatial dependence is displayed.

4.2.2 Visual Representation of the Health Data

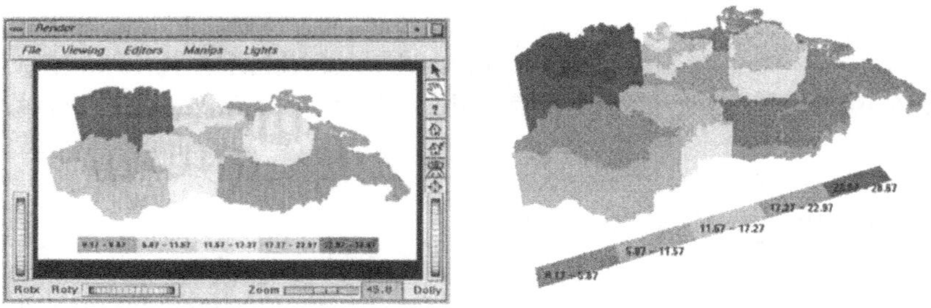

Figure 5: 3D colour coding

Several visualization techniques have been implemented for the graphical representation of the health data, e.g. colour coding, different icons and Parallel Coordinates. A simple colour coding of the areas on the map is used for the display of only one parameter. Since we use 5 colour classes for easier detection of significant areas a 3-dimensional colour coding display can be used where the exact parameter value is encoded in the height (see Fig. 5).

For the graphical representation of several parameters icons are used for displaying either one parameter at different time steps or different parameters at a fixed time . The user can select the icon type, i.e. bar, bar-star, star, pie chart or cross icon, encoding type, i.e. encode parameter in colour or size of icon, and icon size interactively. By clicking at an icon the exact parameter value is displayed in an extra window. (see Fig. 6) shows a *triangle icon.* Triangles are grouped together (in this case 4) to represent the data of different diseases of a fixed time step. Color Coding is used to distinguish the diseases. Furthermore the size of the triangles codes the correspondent quantities. So we can recognize in an intuitive way which diseases occur in a high quantity and which do not. In Fig. 6 the values of the clicked district *Guestrow* are shown on the left side.

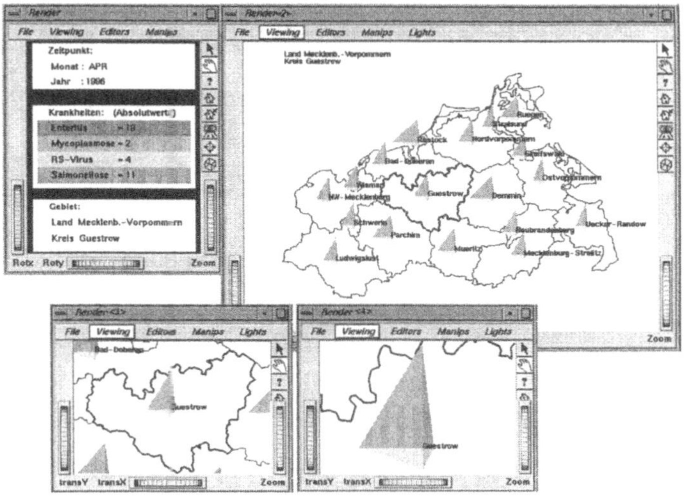

Figure 6: display of the health data with icons

For the display of the temporal course of one parameter we use special icons two most techniques multiple window and animation. A multiple window display is suitable for comparing the data simultaneously in one picture. Furthermore, an animation of a colour coding representation can be used for the display of dynamic courses and changes. The user can select the time interval and the delay time and control the animation interactively.

4.2.3 Realizing further functionality

The event dependent visualization, discussed in 4.1.3, is realized only in a very simple way so far. We use for a threshold-based visualization, e.g. an area is only coloured when the parameter exceds a user defined threshold value.

On the other hand we have implemented the concept of grouping (see section 4.1.3) in MedVis. Geographical areas with the same or similar data or whose graphical representations of the data would be equal are grouped together and are represented by one icon. The groups are formed automatically according to a pre or user defined criterion. In our example (see Fig. 7) the data are represented by bar icons and the geographical areas which belong to the same group are indicated by the same grey colour. Thus the icon represents the data of all areas which have the same grey colour as the area where the icon is positioned at. A further grouping criterion can be defined by allowing only such groups of geographical areas where every area is a neighbour to at least one other area of the same group.

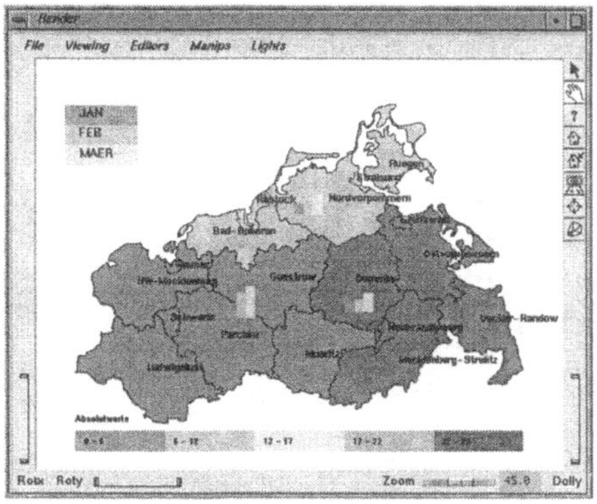

Figure 7: Grouping of geographical areas with similar data

5 Conclusions and future work

To improve the current information flow and the analysis of health data in the public health system, we discussed the visualization of large data sets of multiparameter data with temporal and spatial dependence on different granularity levels. We introduced the concept of the event dependent visualization and the grouping of geographical areas for an intelligent visualization of data by emphasizing relevant information in the graphical representation. Furthermore, we described the prototype visualization system MedVis for the visualization of health data of Mecklenburg-Vorpommern as part of an tele-consultation-system for the monitoring of acute health risk.

The strength of the system lies in great variaty of visualization and interaction functionality and the possibility of generalization to different domain with the task of visualization multiparameter datasets in space and time.
Present limitations result from three reasons:

- The task of contimous data flow at the moment
- The connection to the simulator is not get established (see Fig.1)
- The use of expansive hard- and software.

Therefore future work will focus on improving these data flow (we work on the desing of a data base with a appropiate interface), the connecting with the functionality of MedVis via WWW.
Furhermore, we will look for other approaches for an intelligent visualization of data which allow an easier and faster detection of relevant data features in large multiparameter sets.

Acknowledgements
The authors wish to thank Andrea Schumann for design and implementation work and providing some of the pictures. The work presented is part of the TeCoMed project supported by the German Research Network (DFN).

References

[Dorling94] Dorling, D.: Cartograms for Visualizing Human Geography, in: Hernshaw, H.M.; Unwin,D.J.(eds.): Visualization in Geographical Information Systems, John Wiley & Sons, Chichester u.a. 1994.

[FeBe90] Feiner, S.; Beshers, C.: AutoVisual: Rule based design of interactive multivariate visualizations, IEEE Computer Graphics and Applications

[Gierl96] Gierl, L.; Bull, M.: Architecture of an Early Warning System for Regional Health Risks, Rostocker Informatik Berichte (1996) 19.

[Gierl97] Gierl, L.; Kundt, G. Bull, M.: An Eraly Warning System for detection of outbreaks of epidemics, GEOMED 97, Rostock, Germany 1997.

[Grienstein95] Grienstein,G.: Levkowitz, H.: Perceptual Issues in Visualization, Springer Verlag 1995.

[Inselberg94] Inselberg, A.; Grinstein, G.; Mihalisin, T.; Hinterberger, H.: Visualizing multidimensional (multivariate) data and relations, Hernshaw, H.M.; Unwin, D.J.(eds.): Visualization in Geographical Information Systems. John Wiley & Sons, Chichester u.a.,1994

[Johnson93] Johnson, B.: Visualizing Hierarchical and Categorical Data, Ph.D. Thesis, Department of Computer Science, University of Maryland, 1993

[Liere96] personal talk with Robert van Liere at the 7th Eurographics Workshop, April 1996, Prag

[RDE93] Rew, R.; Davis, G.; Emmerson, S.: NetCDF User's Guide. An Interface for Data Access, University Corporation for Atmospheric Research, 1993

[Schumann96] Schumann, H.; López de Chávez, N.; Graw, K.U.: Visual representation of multiparameter data with spatial dependence, 7th Eurographics Workshop on Visualization in Scientific Computing, Prague, April 23-25, 1996.

[Theisel95] Theisel, H.: Analyse und Visualisierungshilfe fuer mehrdimensionale wissenschaftliche Daten, Informatik - Forschung und Entwicklung, Heidelberg, 1995,10

[WL93] van Wijk, J. J, van Liere R. D.: Hyperslice, Visualization '93, san Jose, ca, 1993.

Content Analysis and Visualization of Epidemiological Documents on the Internet

Bernd W. Kolpatzik, Leo Pfefferer, Albert Schappert

Siemens AG
ZT IK 1
Munich, Germany
e-mail: Bernd.Kolpatzik@mchp.siemens.de

Abstract

Due to the dramatic growth of available data in science, engineering and economics, it has become increasingly difficult to access relevant information and to recognize hidden structures within large data sets. Current approaches for data mining and information visualization are often lacking the necessary efficiency for practical application.

In this paper we propose a framework to structure, visualize and retrieve documents. We present components of the tool-set InfoSphere to analyze documents and to determine their relevance with respect to a set of user defined interest profiles. In a second step, we apply the Siemens visualization tool KOAN (context analyzer) to convert relevance values between documents and profiles into a 3D-representation. KOAN is as an easy-to-use, visual tool, for analyzing data sets, complex relationships and interrelated systems.

We apply our system to structure and visualize a set of internet documents from an epidemiological database. After comparing all documents to a set of user defined profiles, we represent the documents as objects and the profiles as attributes, defining a multi-valued formal context. We explain the transformation of the context into a virtual 3D-space and present the epidemiological documents as a 3D-visualization. With groups of related documents clustered around profile themes or key words, the user may navigate through the information, access documents by topic, and develop an understanding for structure and coincidences of themes.

1 Introduction

Every six years the knowledge of the world doubles. In particular, the explosion of information and data for science, engineering and economics has made it increasingly difficult to sort through available information sources and to retrieve relevant information. Current tools for searching the internet or databases typically provide, unstructured text or long lists of references. More efficient methods for accessing and retrieving information are needed.

Recent literature shows an increasing interest in visualization techniques as means to handle large amounts of information. Wise et al. [WTP 95] describe two techniques to graphically represent relationships between themes and keywords of documents. The first approach displays a set of documents as a two dimensional scatter plot and places documents of similar content near each other. In a second approach the authors present a 3D-landscape, to reflect frequency of recurring themes within a set of documents as mountains and valleys. Hendley et al. [HDW 95] describe a method to display a set of web pages and their links as clusters in a 3D-space. While individual documents repel each other, their web-links or other relationships provide a countering attractive force. An application of information visualization techniques for business application may be found in [SGE 97].

Other approaches to visualizing information are published, and papers are also available on accessing and structuring information. However, the system integration of components for accessing, structuring, visualizing and retrieval is a topic of research.

In this paper we propose a framework to structure, visualize and retrieve documents. We present components of the tool-set InfoSphere to analyze documents and to determine their relevance with respect to a set of user defined interest profiles. In a second step, we apply the Siemens visualization tool KOAN (context analyzer) to convert relevance values between documents and profiles into a 3D-representation.

We apply our system to a set of internet documents, containing epidemiological information of France. Documents in this data base are published about every two weeks and contain dates, information on illnesses, location and impact of outbreaks, forecasts and information on required health care resources. Our method allows the user to easily visualize structures, and coincidences contained in the documents, to navigate through the presented information and to interactively retrieve documents of interest.

In Section 2, we will describe our method to structure sets of documents, to compute similarities between documents, and to obtain the relevance of a document with respect to a profile. Section 3 explains the principles and architecture of the visualization tool KOAN, and in Section 4 we present our results of applying our system to the set of epidemological internet documents.

2 Structuring a Set of Documents

2.1 Estimating Similarity Between Documents

Given a set of documents, we apply a known vector based structuring method described for instance by Salton [Sal 83]. Using simple word statistics, this method estimates content similarity between pairs of documents, which has proven useful for information retrieval applications. This approach has been implemented as part of a framework called InfoSphere, and we apply InfoSphere for content analysis of our document set. Before discussing functions of InfoSphere, we start with a brief review of the notation and the underlying algorithm.

We represent document i as vector D_i, by computing a histogram over all occurring words. Each new word specifies a vector dimension, and the numbers of occurrence are interpreted as the vector elements. Consider the document,

{Influenza Report: Large outbreak of influenza reaches Paris.}

The dimensions of the document are: „influenza“, „large“, „outbreak“, „paris“, „reaches“, „report“. The corresponding document vector,

$$D_1=\{2, 1, 1, 1, 1, 1\},$$

reflects the histogram of words in the document. Note that two different documents will generally have different dimensions.

Similarly, we also define profile vectors, which contain keywords describing the same topic, or related themes. Weighting coefficients for each term compose the elements of a profile vector. Profile vectors can be generated automatically, or can be user specific. We will use profiles to structure sets of documents.

To compute a measure of similarity between two documents, we first project both document vectors onto a common subspace and then define similarity S as the cosine of the angle between the projected document vectors,

$$S(D_1,D_2) = \langle D_1,D_2\rangle/(||D_1||*||D_2||),$$

where $\langle *,* \rangle$ is the scalar product between vectors and $|| \, . \, ||$ defines the euclidean norm of the vector.

This mechanism for computing similarity is rather crude. Ordinary texts consist to a large part of words, which are semantically irrelevant, and the measure can be improved in several ways, by reducing this overhead.

- First, we may eliminate stop-words, which generally depend on the domain of discourse of the particular document space. In many cases, conjunctions, articles and prepositions can be safely removed from the text without changing its content. Removing additional, domain-dependent words may further improve the significance of the measure.

- Second, our similarity metric can be focused on important aspects of the application domain. In this case, we consider only a few concepts or topics, described by words from domain specific thesauri or ontologies.

- Third, we may increase the discrimination power of the metric by applying the so called idf (inverse-document-frequency) correction. In this method, we adjust the weights of words occurring in many of our documents by a logarithmic factor F:

$$\text{IDF\_factor} = \log(\text{number of documents D containing word } w_j \,/\, \text{total number of documents}).$$

Thus, the final measure of similarity includes IDF weighting and is defined over a document subspace, which is obtained after removing stop-words and accounting for restrictions by thesauri.

In the same way as for the similarity measure between documents, we may also compute a metric of relevance for a document with respect to an interest profile. Interest profiles (or profiles, for short,) capture and represent specific queries to the information base or long term information interests.

2.2 Functionality of InfoSphere

We will assume that our documents are already available in electronic form and InfoSphere performs the calculation of metrics introduced above. The system is based on an internal data representation and information objects. An internal repository contains three different aspects of an information object: The data source, a text only representation and a set of annotations. Furthermore, a set of profiles may be specified. Information objects are compared with profiles via the vector-space approach explained above. The computed relevance values of this comparison are stored as a result file. InfoSphere can be instructed to compare a single profile with a list of information objects in a single process.

InfoSphere allows us to test different approaches for structuring information:

1. We may calculate relevance values between documents and profiles directly or
2. we may apply InfoSphere to compute similarities between two documents by treating one document as a profile and computing the relevance of the second document with respect to the first.

Both can be refined by the application of stop-lists and inverse-document-frequency.

Through a set of loosely integrated components, InfoSphere also allows us to automatically access data sources such as a data base or the internet to retrieve documents.

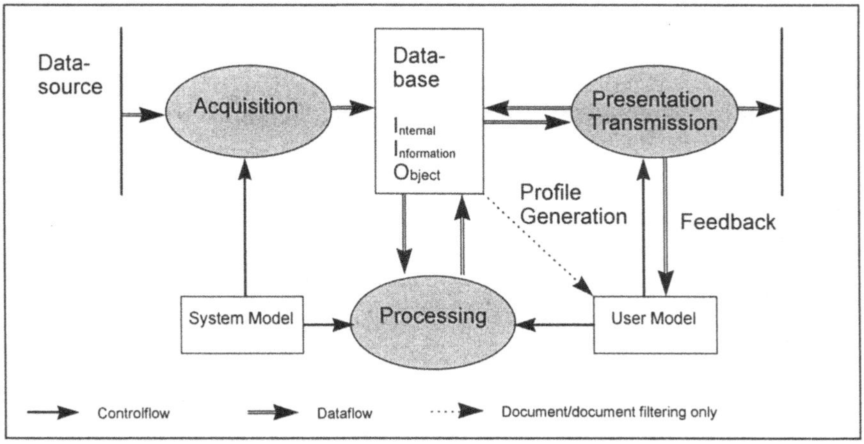

Fig. 1. Illustration of the InfoSphere architecture.

Figure 1 illustrates the architecture of InfoSphere. The acquisition module fetches data and transforms it into the internal format. In the next step a set of profiles has to be generated. The processing module calculates the relevance metrics between documents and profiles and the results are passed on for further processing and visualization.

3 Visualization and Retrieval with Context Analyzer (KOAN)

A key challenge of information processing is to develop intuitive and easy-to-use visualization paradigms [P-S 97], which assist the user to discover hidden relationships and structures in very large data sets. We developed the visualization and data mining tool KOAN (context analyzer, german: '*Ko*ntext-*An*alysator') [Pfe 96, PFP 96]. KOAN is designed to support the ability of the human observer to orient and navigate naturally through 3D-space, and we apply this tool to visualize and retrieve our set of internet documents.

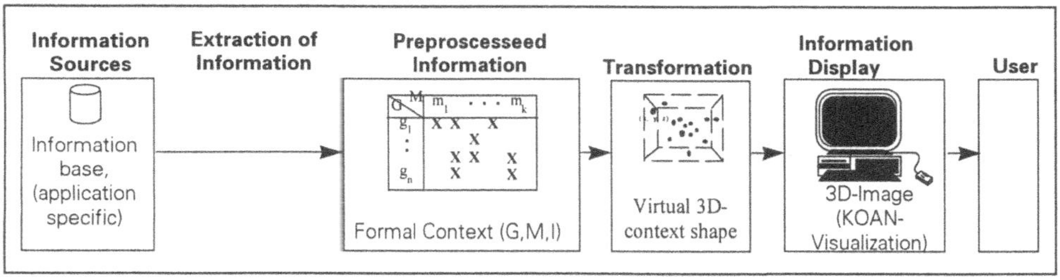

Fig. 2. Illustration of the system architecture of context analyzer KOAN.

Figure 2 illustrates the system architecture of KOAN. After accessing data from a given information source, we preprocess the information and apply the concepts of a formal context. Formal contexts describe information as objects and attributes. We determine the attributes, which are associated with each object, and we obtain a set of relations to form an object-attribute or incidence matrix.

Objects, attributes and incidences generally span a multi-dimensional context space and are not easy to visualize. However, applying the idea of representing content similarities by spatial proximity allows us to map information objects and their attributes as clusters into a 3D-space (information space). The approximate embedding of the formal context in 3D-space is achieved by an iterative non-linear optimization method, with respect to a set of spatial constraints [FJM 94].

After computing an optimized 3D-representation of our data, we project and display the information objects and attributes as color coded geometrical symbols on the computer monitor. We also implemented interactive controls to move the user's viewpoint relative to the data for choosing the optimal presentation of the spatial structures and giving the impression of navigating the 3D-information space. Selecting a symbol, retrieves and opens the associated information object or attribute, for example as a web-page in an internet browser.

Displaying information in 3D-space corresponds to a familiar environment and supports the interpretation procedures of human beings to recognize groups of spatially neighboring objects as related items. This intuitive method of visualization allows the user to rapidly scan large volumes of data and memorize its correlation. KOAN also provides access to details and selected information when needed. Structures such as coincidences of attributes or similar themes between documents are reflected by spatial proximity and are easy to recognize. Comparisons between structures of several sets of data are also easily accomplished.

Applications for contextual information processing range from archiving, operating power plant control rooms to software reengineering and to data mining.

4 Visualization of Internet Documents

We have tested our information system on a set of internet documents from the epidemiological data base „Sentiweb"[1] of the French General Practitioners. These documents are published every week and contain dates, information on illnesses, location and impact of current outbreaks, forecasts and information on required health care resources in France.

For our experiments, we define twenty interest profiles, containing keywords with weighting coefficients in four categories: Illnesses (8), seasons (4), general medical terms (4) and geographic regions (4). Table 1 lists sample profile for each category. We apply InfoSphere to retrieve and to compute relevance values for a set of 84 documents with respect to the profiles. The relevance values are converted into a 3D-representation, applying the concepts of a formal context.

[1] http://www.b3e.jussieu.fr:80/sentiweb

Illnesses Symptoms Circumstances Synonyms (8)	Seasons Months (4)	General Terms (4)	Geography Regions (4)
flu or ILI 1.0	winter or holiday or Christmas or Dec or Jan or Feb 1.0	Practioner 1.0	centre or center or paris or PACA 1.0
diarrhea 1.0 accute 0.5		serological 1.0 salivary 1.0 HIV 1.0 test 1.0	

Table 1. Sample profiles in four categories. Each profile consists of one or more weighted terms and boolean operators.

Figure 3a shows the positions of the profiles (spheres) in 3D-space. Profiles which are positioned in spatial proximity, are relevant to similar subsets of documents. The user can select among the presented profiles, and locate related documents. Figure 3b displays two selected profiles („fever“ and „summer“) with their associated documents (cubes). Selecting one of these documents as illustrated in Figure 3c brings up all related profiles to the document (G10).

Utilizing this tool, the user can determine common topics for a group of documents, topics related to a given document, or find groups of related topics or related documents.

Alternatively, we may also obtain profiles based on the documents alone. As described in Section 2.2, InfoSphere allows us to compare a set of documents directly with each other. To reduce overhead and to obtain more meaningful clustering results, we first passed our document set through a word list of medical terms and then compared the reduced documents directly with each other. Figure 3d illustrates the resulting clusters of documents. Recall that spacial proximity corresponds to similarity in content.

Our method allows the user to easily visualize structures, and coincidences hidden in the documents, to navigate through the presented information and to interactively retrieve documents of interest. This representation enables the user to quickly locate all documents related to a specific illness, or to examine relationships and coincidences between illnesses and other circumstances, such as the season of the year.

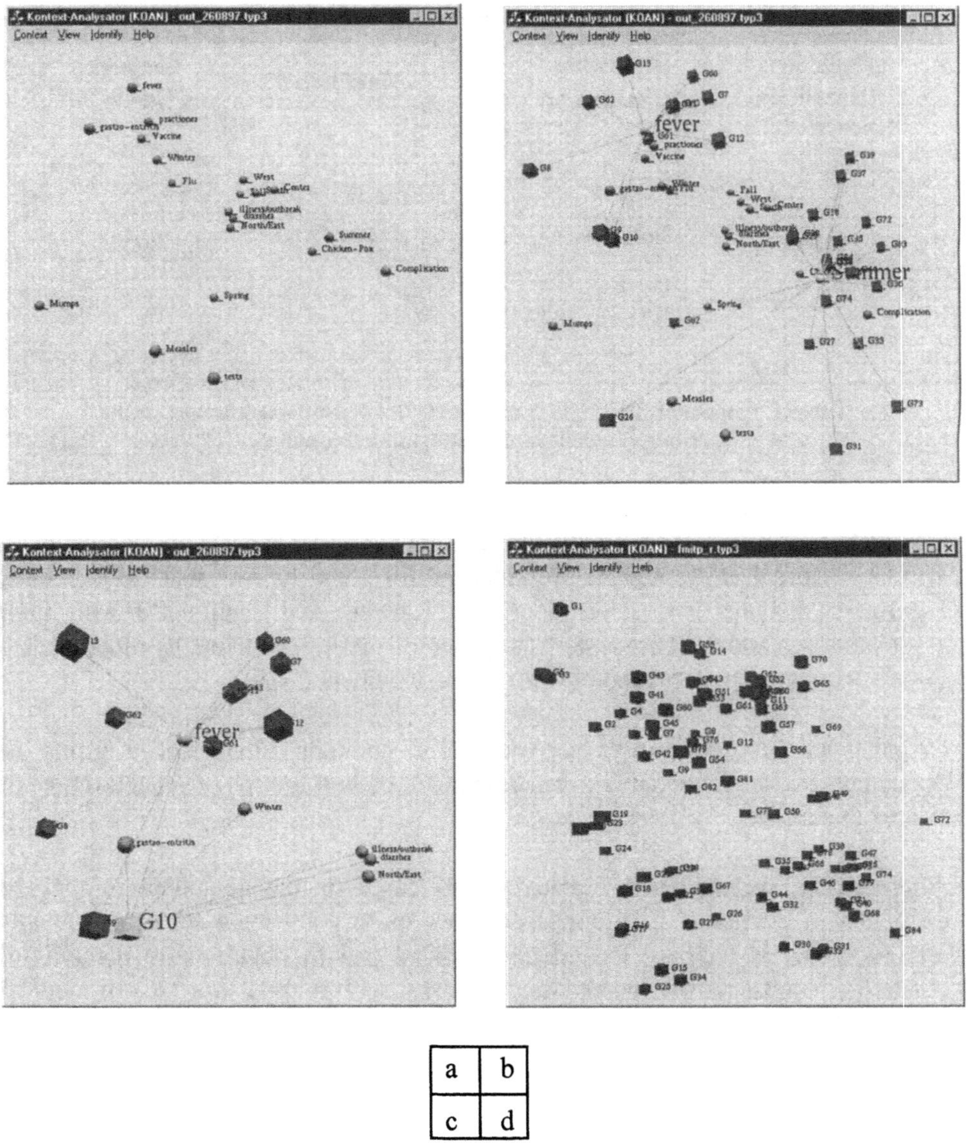

Fig. 3. View of epidemiological documents. Spatial proximity in 3D-space indicates similarity in content. a) Interest profiles (spheres) only. b) Illustration of two clusters of documents (cubes) associated with two profiles. c) Detail, illustrating the highlighting function (profile fever) and one selected document (G10) with its associated profiles. d) Documents are passed through word list and compared to each other (no profiles).

In a recent version, we are using internet technology (JAVA, VRML2.0) to display the information space and to enable the user to bring up documents of interest in an HTML frame within a standard internet browser.

5 Summary

In this work, we have presented an integrated system which enables the user to access, structure, visualize and interactively retrieve documents from an information base and applied our system to a set of epidemiological documents from the internet. The integration of document analysis, visual presentation and graphical interface enables the user to efficiently obtain information related to a topic and to discover structures and hidden relationships within the document set.

References

[WTP 95] J. Wise, J. Thomas, K. Pennock, D. Lantrip, M. Pottier, A. Schur, C. Vern, „Visualizing the Non-Visual: Spatial analysis and interaction with information from text documents", *IEEE Symposium on Information Visualization, 1995.*

[HDW 95] R. Hendley, N. Drew, A. Wood, R. Beale, „Case Study Narcissus: Visualizing Information", *IEEE Symposium on Information Visualization, 1995.*

[SGE97] T. C. Sprenger, M. H. Gross, A. Eggenberger, M. Kaufmann: A Framework for Physically-Based Information Visualization. Appeared in: Visualization in Scientific Computing '97, *ISBN 3-211-83049-2, Springer Verlag, 1997*

[FJM 94] T. Führing, K. Jacoby, R. Michelis, J. Panyr, „Kontextgestaltgebung: Eine Metapher zur Visualisierung von und Interaktion mit komplexen Wissensbeständen", *Nachtrag Schriften zur Informationswissenschaft Bd. 16; von Rauch, W.; Strohmeier, F.; Hiller, H.; Schlögl, C. (Hrsg.): Mehrwert von Information: Professionalisierung der Informationsarbeit (Proceedings des 4. Internationalen Symposiums für Informationswissenschaften (ISI '94) in Graz, Österreich, 2.-4. November 1994). Konstanz : Universitätsverl. Konstanz (UVK), 1994*

[P-S 97] L. Pfefferer, D. Schütt, „InfoSphere™-V: A New Approach to 3D-Visualization of Information", *to appear at Springer Publications, Heidelberg 1997.*

[Pfe 96] L. Pfefferer, „Objektzentrierte Visualisierung mehrdimensionaler Daten als Erweiterung konventioneller Datenbankmodelle“, *München: Herbert Utz Verlag, Dissertation, 1996.*

[PFP 96] J. Panyr, T. Führing, U. Preiser, „Kontextuelle Visualisierung von Informationen“, *Proceedings des 19. Oberhofer Kolloquiums über Information und Dokumentation, Oberhof, 18.-20. April 1996, S. 217-228*

[Sal 83] G. Salton, M. J. McGill, „Introduction to Modern Information Retrieval“, *McGraw-Hill Book Company, 1983*

A NEURAL NETWORK MODEL FOR THE SPREAD OF COMMUNICABLE DISEASES

Laurent Toubiana and Jean-François Vibert

B3E, INSERM U263, ISARS
Faculté de médecine Saint-Antoine
27, rue Chaligny
75571 Paris Cedex 12, France.
Tel: 44 73 84 52
Fax: 44 73 84 62
e-mail: toubiana@b3e.jussieu.fr

Abstract. Since 1984, the French Communicable Disease Network (FCDN) has collected data on eight communicable diseases. This has allowed the investigation of the complex spatio-temporal dynamics of influenza-like diseases.
We propose a model based on a neural network analogy to simulate the propagation of such epidemics. We simulated the spatio-temporal evolution of influenza epidemic which corresponds qualitatively to the actual data observed by the FCDN.

1 Introduction

Some viruses spread by using the individuals they infect. Consequently, the movement of individuals permits the viability of viruses. The dynamics of communicable diseases can be described as a complex phenomenon: the diffusion in space and time of virus populations on other populations (for example human).
In a population, each infected individual transmits a virus to other individuals he has contact with. These latter individuals may develop the disease, depending on their susceptibility, and transmit the virus to others.
Individuals may be infected or not, and thus they behave in a non linear way. Individuals thus constitute a network of connected non linear entities exchanging virus. This is rather similar to neurons, that are also non linear networked objects, exchanging information inside neural networks. The behavior of such networks is complex and has been thoroughly studied during the last decade, in both fields of artificial neural networks and of computational neurosciences.
The analogy between humans exchanging viruses and neurons exchanging information is rather straightforward. In a neural network, each neuron integrates the inputs it receives, and if a threshold (the non linearity) is reached, an action

potential is fired and transmitted with a delay to other neurons it connects. Then the firing neuron becomes less receptive for a short time, called the refractory period. It could be interesting to benefit from the large amount of work done on neural networks to understand the collective behavior of neurons when included in a network, since the spread of epidemics can represent a similar problem. In the analogy proposed, humans infect other humans if they are susceptible. If they are infected they get some immunity leading to less susceptibility. This is quite similar to neurons connected by excitatory connections to other neurons in a neural network.

Thus, it is possible to consider a territory as a conducting medium containing « excitable » elements which are contaminated and contagious in the excitable state. When a neuron fires, it transmits its action potential to the other neurons it is connected with. If the target neurons receive enough action potentials, it fires also, and can propagate the signal inside the network. Excitatory neural networks were studied by several authors, since they present interesting properties for the brain information processing [1,2]. Among other properties, excitatory neural networks can display recurrent bursts of activity when there exists an interneural transmission delay [3,4] and a diluted excitatory connection [5]. This is reminiscent of the epidemics recurrence observed by the "Réseau Sentinelles", the so called French Communicable Disease Network (FCDN). The FCDN collects data in France from eight infectious diseases, some since 1984, directly from French general practitioners using the Minitel network [6]. From these data, temporal series as well as spatial representations (maps of epidemics in France) can be computed [7] and show, for some diseases such as influenza, the progressive propagation of recurrent epidemics in the French territory with a one year approximate periodicity.

Taking advantage of the experience we gathered in computational neuroscience, we developed a connectionist model based on the neuron analogy to describe the dynamics of communicable diseases in human populations. The model consists of elementary interconnected units constituting a network. Each unit presents a non-linear response to the inputs it receives. This work mainly uses methods and tools developed or derived from methods for computational neuroscience [8] to build a model supposed to mimic the influenza epidemics invading France each winter.

As a first stage, this paper is a feasibility study and presents qualitative results obtained by using the metaphor proposed here.

2 The model

The studied network is made of a collection of interconnected elementary units. Each unit can be viewed as representing a population of individuals, just like the graded response model of neuron used in some neural networks is viewed as representing a population of neurons [9]. These units are connected together to constitute a network in which an epidemic can spread.

2.1 The unitary element

The elementary structure of the model is a non linear object defined by its state *e(t)* and its threshold *s(t)*. The state *e* represents the infection incidence and the threshold *s* the susceptibility level. At rest, the system is in a steady state where the state *e* is below the resting threshold *S*. When a part of the population get infected, the state *e* increases, and when it reaches the threshold *S*, it increases rapidly (an epidemic begins) and the threshold increases also, accounting for the decrease of susceptibility of the population due to the induced immunity. Both variables are thus interdependent and time related.
Each unit receives inputs from other units. These inputs are weighted by a factor called connection weight ω_{ji}, representing the influence of the inputs from unit *i* on this specific unit *j*. It can be seen as proportional to the number of contacts allowing infection transmission between two populations. Transmission can be delayed by a delay D_l, representing the incubation period. For this biological system some noise ε has been introduced.
The temporal evolution dynamics of the state $e_i(t)$ and of the threshold $s_i(t)$ of the element *i* are described by the following set of delay equations :

$$e_i(t) = \Sigma_{j \neq i}\ \omega_{ji}\, e_j(t-1) + \varepsilon(t)$$
$$s_i(t) = S_i$$

If $\quad e_i(t) > s_i(t)$

$$t_{io} = t$$
$$e_i(t) = e_i(t) + A_{it_{io}}\, e^{-(\mu_{itio} - t)^2 / 2\, \sigma_{itio}^2}$$
$$s_i(t) = A_{it_{io}}\, e^{-(\mu_{itio} - t)^2 / 2\, \sigma'^2_{itio}}$$

where

t_{io} = infection time for unit *i*

ω_{ji} = connection weight from *i* to *j*

Si = resting threshold value (sensitivity threshold)

$\varepsilon(t)$ = noise

$A_{it_{io}} = f(A_i,\ e_i(t_{io}),\ e_i(t_{io}-1))$ which is the maximum value of incidence during an epidemic period for unit i computed at infection time (t_{io}). We cannot forecast this value before the epidemic began. The assumption is made that it depends on the difference between incidence at infection time and incidence just before this date. Simulation are made using :

$$A_{it_{io}} = \frac{A_i}{1 + e^{-10\left(\frac{2(e_i(t_{io}-1) - e_i(t_{io})) - 1}{2A_i}\right)}}$$

$$\mu_{itio} = t_{io} + D_l + D_e/2$$

$\sigma_{itio} = D_e/2$

$\sigma'_{itio} = D_r/8$

A_i = maximum value of incidence during an epidemic period
D_I = delay corresponding to incubation period
D_e = duration of epidemic period
D_r = nominal value of immune period duration (refractory period)

Figure 1 shows simultaneously the temporal activity of both the state variable *e(t)* and the threshold variable *s(t)*, as well as the dynamics of their non-linear relationship. Each element reacts when the received signal exceeds the sensitivity threshold. The state variable *e* increases sharply after the period D_I (corresponding to the incubation delay), corresponding to the start of an epidemics. The threshold *s* increases also, corresponding to the start of a refractory period of lesser susceptibility. Both variables return back to their rest value with different time constants respectively, D_e corresponding to the duration of a local epidemic period and D_r corresponding to the duration of the immune period (refractory period). For sake of brevity, we will call the element an "epineuron" because it represents the epidemics as a neuron.

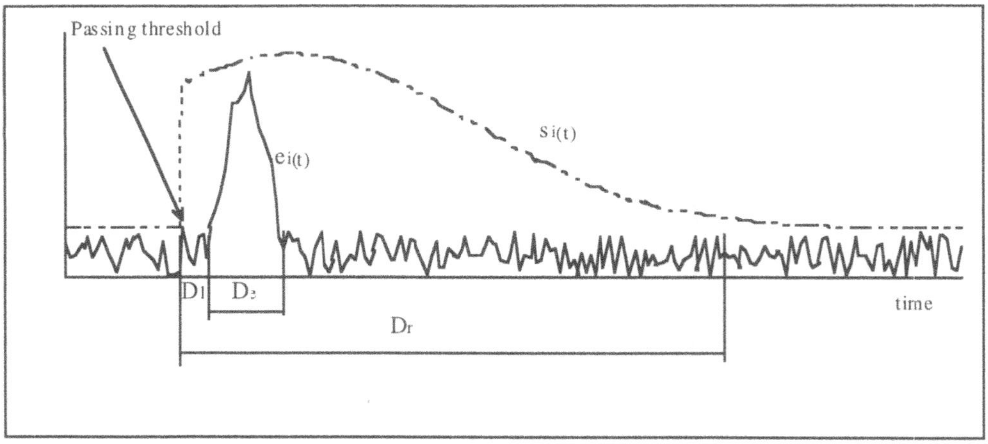

Fig. 1. Dynamics of an epineuron's state and of its threshold.
Each epineuron reacts after a delay Dl when the received signal exceeds the sensitivity threshold Si(t) (dashed line). The threshold evolution is a function of the epineuron's state. The threshold increase represents the refractory period of lesser susceptibility. Abscise: time in weeks; ordinates: incidence in arbitrary units.

2.2 The network

Epineurons defined above can transmit their state to other epineurons they connect. A connected set of epineurons constitutes a network where each epineuron may

receive inputs from which it computes its own state, and if this new state overtakes threshold, the threshold increases, as seen previously and an output is transmitted to the epineurons it has connection to.
For each connection a coefficient defining the coupling strength (or connection weight as stated above) between the epineurons is attributed when the network is defined. In this paper, epineurons are supposed to represent populations, thus the connections weight will represent the interaction coefficients between these populations, and here more precisely these coefficients will be computed according to the spatial (geographical) location of the population as explained below.

2.3 Application to influenza epidemic

Influenza is one of the eight communicable diseases monitored by the French Communicable Disease Network (FCDN) since 1984 [10]. It is an acute disease which affects the respiratory system. Recovery occurs spontaneously within two to seven days. The incubation period is short, 24 to 72 hours. The period in which an individual is contagious is thought to be less than three days. The virus is transmitted by droplets named "Flügge", which are exhaled by the infected person and consequently inhaled [11]. We use actual data collected by the FCDN of the influenza morbidity from 1984 to present.

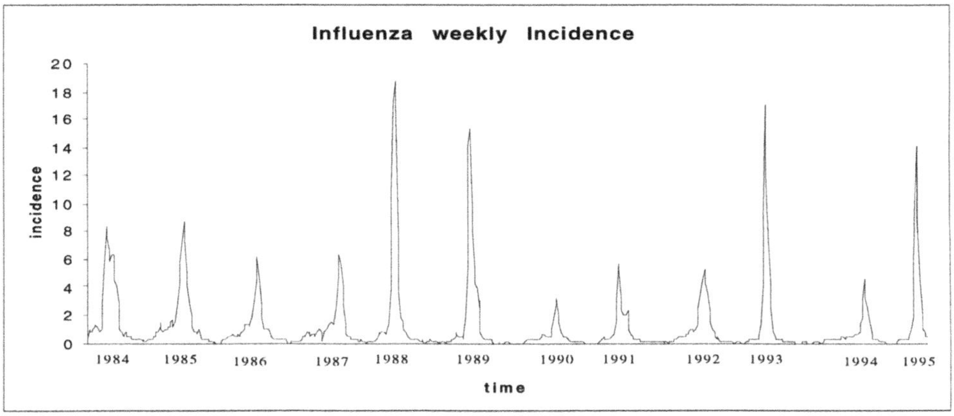

Fig. 2. Temporal dynamic of the incidence of influenza from 1988 to 1995. Each spike corresponds to an epidemic period.

Figure 2 shows the increase in the incidence of influenza during the winter season of each year. This corresponds to the influenza epidemic period that lasts approximately six weeks and can reach an estimated total number of one to four million cases.

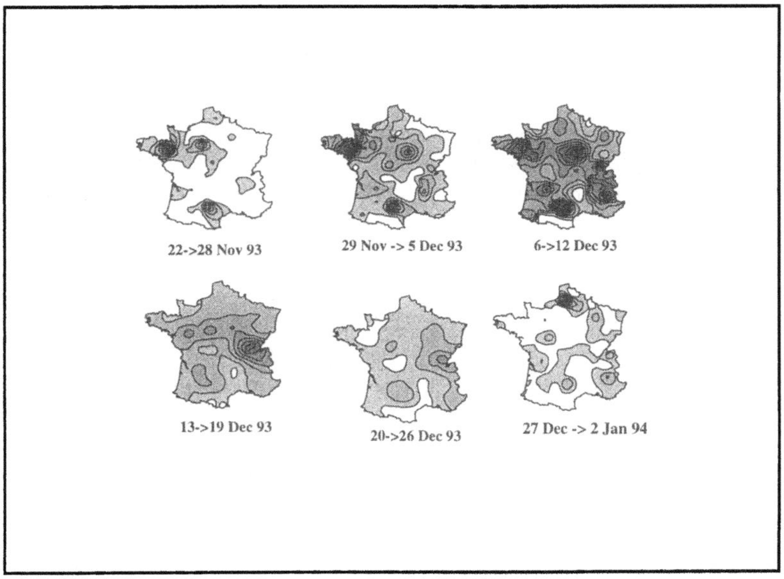

Fig. 3. Spread of the influenza epidemic during winters 1993/94 produced with SITIE

Figure 3 presents typical maps of influenza incidence as produced weekly by FCDN which shows the complexity of this spatio-temporal phenomenon.

2.4 Modeling spatial propagation of an epidemic

It is possible to adjust model parameters based on the influenza morbidity data in order to perform simulations of an epidemic.
The French territory is partitioned into 94 subspaces corresponding to the 94 French departments (administration definition). Any region may receive individuals from any other regions. Knowing the epidemic state of the region of origin, it is possible to compute for each region the influence of people migration regarding the epidemics.
It is difficult to obtain data on the population flow between departments. A gravitational equation [12,13,14] was used to establish the connection weights interaction coefficients (coupling) between spatial entities (here the departments). This equation allows to evaluate migration flows from data of spatial location and demography. ω_{ij} is proportional to the number of persons coming from department i to department j.

$$\omega_{ij} = P_j/(d_{ij})^2 \,/\, \Sigma_{k \neq i}^{n} P_k/(d_{ik})^2$$

with $\Sigma_j \, \omega_{ij} = 1$

n is the number of spatial entities

d_{ij} is the Euclidean distance between i and j
P is the population of the spatial entity (here the department).
Each department is represented by one of the epineurons defined above. Connection weights ω_{ij} correspond to coupling coefficients. The epineuron's state corresponds to the disease incidence in the corresponding department.

Any region may receive individuals from other regions. Knowing the state of the epidemic in the region of origin, it is possible to compute for each region the influence of migration regarding the epidemic. When the ratio of total number of infected individuals to the total number of individuals in the considered region passes a threshold, an epidemic begins.

At each time step, the weighed sum of incoming states is computed for each epineuron. When this sum exceeds a threshold, the epineuron is "activated", and its output propagated to other epineurons. The network activity will be estimated by the sum of the state variable e of the constituting epineurons at each time t. This indicator makes sense, since it represents the temporal evolution of the total incidence of a given infection among the whole network.

3 Results

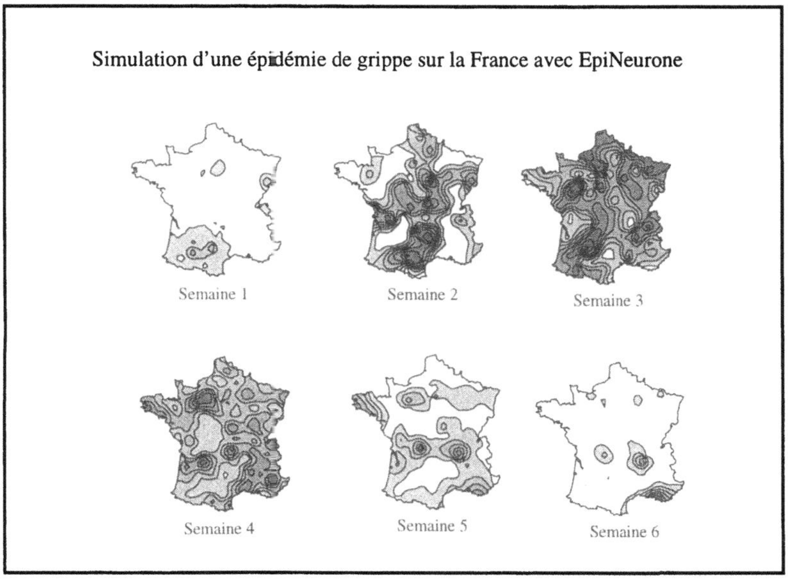

Fig. 4. Simulated spread of an influenza epidemic

Figure 4 presents a series of maps resulting from the simulation of a fully connected network consisting of 94 elements (French departments). The weight of each connection was computed as stated above. Initial conditions were generated using a gaussian distribution (*m=1, sd=2*). The threshold value was set to three. A gaussian noise (*m=0, sd=1*) was added to the state of each element. Figure 5 shows the result from a 500 weeks simulation, with a one week simulation step.

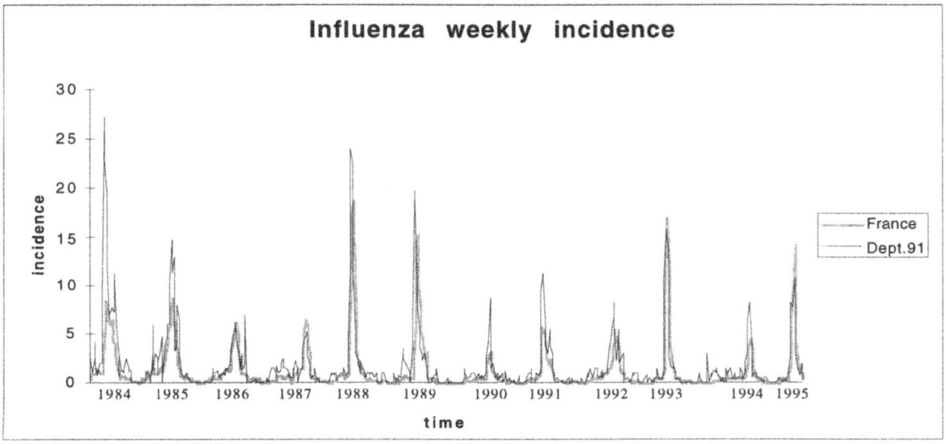

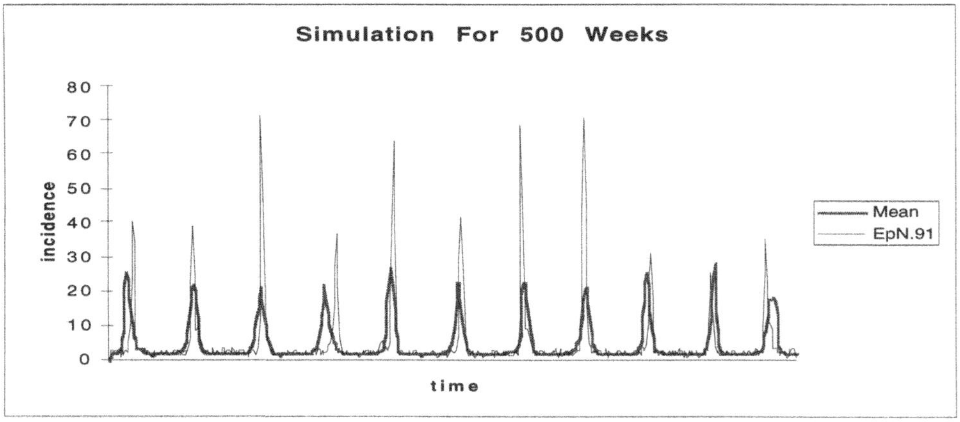

Fig. 5. Long term temporal evolution of a simulated epidemic. Top panel: Observation of actual influenza incidence between 1984 an 1995. Same data as in Fig 2. Bottom panel: Simulation of a network of 94 epineurons during 500 weeks. The curves show the evolution of the incidence of influenza. Compare both curves. Thin line: one department. Thick line mean incidence in the whole country. Abscissa: time in weeks; ordinates: incidence in cases for 100,000 habitants.

After a few simulation steps some elements are active (infected) due to initial conditions and added noise. Then, the network interaction causes the propagation of the "epidemic" to all of the French territory. For each infected element, the threshold increases, causing a refractory period. Consequently, these element cannot be reinfected during the current epidemic that will naturally vanish. The network has a low level of incidence during the refractory period of the elements. After this period a new activity cycle can begin. Figure 4 presents maps from a simulation of the model. The spatio-temporal evolution of this phenomenon corresponds qualitatively to the actual evolution of the epidemic as observed by the FCDN shown in figure 3.

4 Discussion

This paper proposes a method for investigating the spread of an epidemic using neural networks analogy applied to actual epidemic data. Several mathematical and statistical models [15,16,17,18] were developed to describe the spread of an epidemic based on epidemiological data. Models were proposed to simulate the interactions of populations in the spatial domain. Some were based on spatial Volterra-Lotka equations [19], others on multi-agents systems and distributed artificial intelligence [20]. Artificial neural network methods have been applied to epidemiology [21] in a different way, since they don't use neural network as a model of interconnected entities, but as a tool to analyze epidemiological data. As far as we know an approach similar to those proposed here was also proposed by Christea and Zaharia [22,23] but neurons represented individuals, not populations and simulations were done with fictive data and not analyzed in the spatio-temporal domain. They also showed incidence increase followed by a vanishing, as observed here. The connectionnist model proposed here allows to generate population behaviors at global level. In this model, a behavior law is not defined for the entire population but only some fundamental properties of elements constituting this population, and gets an emergent result at a collective level.
In conclusion, the neural metaphor in which is rooted this work leads to interesting results since it allows to propose a mechanism to explain the recurrence of incidence peaks and their relatively well organized geographical propagation. Our aim is to further develop this model as a prediction tool and as an alert system [24] for the epidemologists.

5 References

[1] MacGregor RJ and Palasek RL : Computer simulation of rythmic oscillations in neuron pools. Kybernetick 16, 79-86 (1974).

[2] Amari S : A method of statistical neurodynamics. Kybernetik 14,201-215 (1974).

[3] Vibert J-F, Pakdaman K, Azmy N : Inter-neural delay modification synchronizes biologically plausible neural networks. Neural Networks, 7 : 589-607 (1994).

[4] Pakdaman K, Vibert J-F : Modeling excitatory Networks. J SICE, 34:901-906 (1995).

[5] Pham J, Pakdaman K, Vibert J-F : Simulation of an excitatory network involved in the control of the respiratory rhythm. CNS*1996. Fifth annual Computational Neuroscience Meeting (1996)

[6] Valleron A-J and Garnerin P : Computerized surveillance of communicable diseases in France. CDR Review, 3 : 82-87 (1993)

[7] Toubiana L, Vibert J-F, Garnerin P, Valleron A-J : SITIE : A health care workstation integration archtecture for epidemiologists. Computer and.biomed. Research 28,100-115 (1995)

[8] Vibert J-F, Pakdaman K, Cloppet F, Azmy N : NBC : A Workstation for Biological Neural Network Simulation Ed. Josef Skrzypek 113:133 (1994)

[9] Cowan J, Ermentrout GB : Some aspects of the 'eigenbehavior' of neural nets. In S.A. Levin (Ed.) Studies in Mathematical Biology, part I, 67-117, The Mathematical Association of America.

[10] Valleron, A-J., E. Bouvet, Ph. Garnerin, J. Menares, I. Heard, S.Letrait, J. Lefaucheux.: A computer network for the surveillance of communicable diseases : the French experiment. Am. J. of Pub. Health. 76: 1289-92 (1986).

[11] Cliff AD, Haggett P, Ord JK : Spatial aspects of influenza epidemics. Edited by Pion Limited, London. (1986)

[12] Reilly W : The laws of retail gravitation (1931)

[13] Converse PD : New laws of retail gravitation, The J. Of marketing 14:379-384 (1949)

[14] Wilson AG: A statistical Theory of spatial distribution Models Transportation Research 253-269 (1967)

[15] Anderson RM and May RM : Infectious diseases of Humans. Dynamics and control. Oxford Sciences Publications (1991).

[16] Longini, IM : A.Mathematical model for predicting the geographic spread of new infectious agents. Mathematical biosciences 90:367-383 (1988).

[17] Baroyan OV, Rvachev LA, Basilevsky UV, Ermakov VV, Franck KD, Rvachev MA, Shashkov VA : Computer modelling of influenza epidemics for the whole country (USSR). Adv Appl Prob 3: 224-226 (1971)

[18] Cliff AD and Haggett P : Atlas of Disease Distributions, analytic approaches to epidemiological data, Blackwell Publishers, Oxford (1992).

[19] Dendrinos DS and Mullaly H : Urban Evolution. Studies in the Mathematical Ecology of Cities, Oxford, Oxford University Press (1985).

[20] Ferber J : Des objets aux agents. Université Paris VI, thèse de doctorat (1989).

[21] Quandt K, Waschulzick T, Lewis M, Hormann A, Engelbrecht R, Brauer W : Evaluation of epidemiological data with neural networks, 1993 International Neural Network Society Annual Meeting I-257 I-260 (1993)

[22] Cristea Al, Zaharia CN : Utilization of neurocomputing in the simulation of some viral epidemics. Europ simulat Multiconf Prague.; 784-789 (1995)

[23] Cristea Al, Zaharia CN, Popescu D, Bota M. : Utilization of a neural neutwork to simulate some human behaviors. Europ simulat Symp Istanbul.; 340-345 (1994)

[24] Valleron A-J., Chauvin P., Flahault A., Vibert J.-F. : Les systèmes d'alerte: passé, présent, futur TCB 6:443-454 (1994)

Identifying clusters in spatial area data

Friedrich Gebhardt
GMD – German Research Center for Information Technology
Intitute for Applied Information Technology (FIT)
D-53754 Sankt Augustin
friedrich.gebhardt@gmd.de

Abstract

A task peculiar to *spatial* data is to find spatial clusters. Three alternative tests will be given here for the case that a region is divided into a number of districts and each district is assigned a numerical value, i. e. the value for a variable of interest. A cluster of districts with high (or low) values cannot just be based on the adjacency of districts since then a random assignment would already produce too many and too large clusters. Thus the cluster definition is based on district triplets, essentially three districts meeting at a common corner. Two tests are designed for a division of the area into marked and unmarked districts. The third, more elaborate test uses the given values directly.

1 What is a cluster?

Throughout we consider spatial area data. A region is divided into F non-overlapping districts (plots, fields) and for each district a numerical value is given, usually the value of a variable of interest. The question is: are the districts with high values concentrated or clustered somewhere?

Often there is no rigid boundary for "high" in this sense, so the numerical values have to be taken into account: a small group of districts with very high values may be more important than a larger group with moderately high values.

Sometimes there is a rather clear boundary for "high". Then we have two subsets of the area: the districts with values above and below this boundary. We may just consider the values to be either 1 (high) or 0 (low); so this situation is a special case. It arises directly if one of the two subsets is characterized by a non-numeric property. e. g. election districts won by a certain party, plots of land where a certain plant is growing, statistical districts where a certain disease has been reported.

We will assume that every single district is contiguous: if there exist actually exclaves, they are either neglected or treated as separate districts. On the other hand, it is permitted for a district to have inclaves. i. e. one or more districts totally surrounded by that district. This occurs in administrational districts (such as German counties).

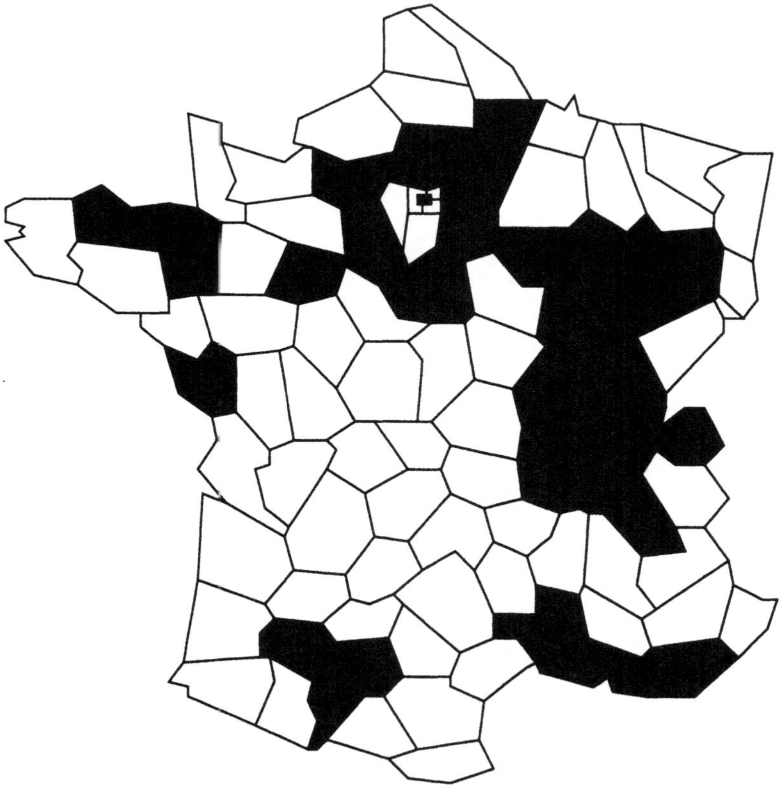

Figure 1: French départements with reported cases of mumps in the first quarter of 1995. Data taken from Le Réseau Sentinelles.

For an example consider figure 1. It shows the French départements (without Corse) where mumps has been reported in the first quarter of 1995. The data are taken from Le Réseau Sentinelles [Toubiana *et al.*, 1995; Toubiana and Flahault, 1998], accessible on the internet at `http://www.b3e.jussieu.fr:80/sentiweb/en/sommaire.html`. Is there a suspicious clustering of départements with reported cases?

What is a cluster? One might try to say "A group of at least k adjacent districts with high values". The shaded region in the east of France together with the extension around the Ile de France appears to constitute a non-random concentration. Another quite similar situation is shown in figure 2 – however, this arrangement of marked districts has been produced by randomly selecting 30 départements out of 94 (without Corse) and it is fairly typical.

This example shows that the marked districts tend to build long chains, which thus are not an indication of clustering. Random just does not mean evenly distributed. In order to identify non-random concentrations we need a stricter notion of clusters leading to more compact regions.

Our definition is based on triplets of districts. A triplet is essentially a set of three neighbouring districts having a corner in common. But sometimes four or more districts meet at a corner (this happens once for the states in the US and often for statistical districts in a city when major streets or railway lines form the boundary). Therefore the definition has to be refined slightly. A *triplet* is a set of three districts such that they have a corner (node) in common and one of them shares with each one of the other two a boundary (edge) having that corner as an endpoint.

A *cluster* is defined as a set of strictly overlapping triplets (that is, the set of triplets cannot be split into two non-overlapping parts). The clusters of interest will have an additional property such as a mean value of the underlying variable sufficiently above average.

In figure 2 we find two clusters, a larger one in the north-east and a small one, consisting just of one triplet, in the east. The dots identify corners where three marked districts meet, i. e. triplets.

For practical purposes, the cluster definition has to be expanded somewhat. It happens in some areas that a district has only one neighbour; thus it could never belong to a cluster. Examples occur in Germany where several urban counties (Stadtkreise) are entirely surrounded by a rural county (Landkreis). Therefore we define:

If a (valid) cluster contains a district A and district B has district A as its only neighbour, then adding B to the cluster gives again a (valid) cluster.

In the following sections we will propose tests for determining spatial clusters but we will refrain from mathematical details that will be published elsewhere, see meanwhile [Gebhardt, 1997b; Gebhardt, 1996; Gebhardt, 1997a].

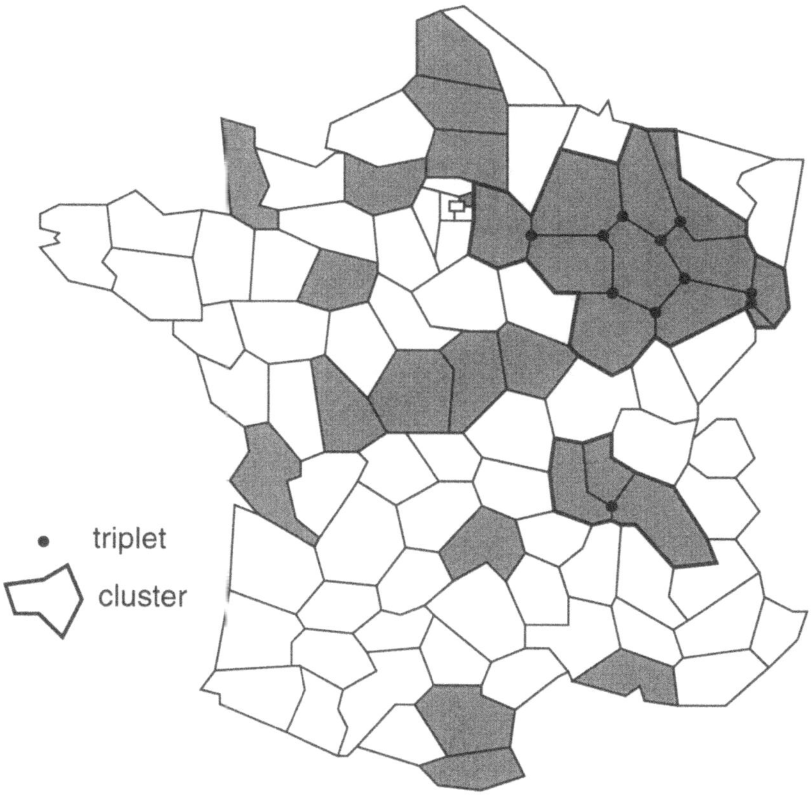

Figure 2: 30 French départements selected at random.

2 Tests for binary data

First we consider the special case of binary data: there are only two admissible values for the districts. So we distinguish the districts by being *marked* or *unmarked* assuming that the marked districts are the minority. Other authors call the districts *black* and *white*.

When a given number M of districts is marked at random in an area, one can find by simulation the statistical distribution of the number M_c of marked districts belonging to a cluster and in particular estimates m for the expectation and s for the standard deviation of M_c. From several simulation experiments one finds that both measures, taken as a function of the number of marked districts, M, can be approximated surprisingly well by polynomials of fourth degree without low-order terms. Thus one need not perform the simulations for all or a large number of values M; a few of them suffice to determine the coefficients in the two polynomials.

The test statistic M_c is an integer; therefore one should apply a continuity correction as usual. As a result one has a statistical test for this area: if

$$M_c > m + \frac{1}{2} + d \cdot s$$

with an appropriate d (perhaps $d = 2$ for a single test, $d = 2.5$ if several tests are being performed and $d = 3$ or even larger in a data mining situation where conceptually hundreds or thousands of tests are made), the arrangement of marked districts is considered clustered rather than random. Conversely, if $M_c < m - \frac{1}{2} + d \cdot s$ for an approppriate negative value of d, the marked districts are significantly unclustered, i. e. too evenly spread out.

Simulations with several areas indicate that the test is sufficiently correct: the one-sided error probabilities for $d = \pm 2$ are actually near 2.5%, those for $d = \pm 3$ between 0 and 0.5% (rather than 0.14% for the normal distribution). Deviations are primarily caused by the test statistic being an integer or by the fact (for larger $|d|$) that 0 and M are natural boundaries for M_c, the number of districts in clusters.

This test is tailor-made for a particular area and insofar no approximation (the approximations for m and s are good enough for practical purposes and taking proper values for d from the experiences with normal or t distributions will also suffice). However, they need a sequence of simulation experiments for each new area.

An alternative test does not need simulations but is somewhat less to the point: it does not test directly for clusters but for adjacencies between marked districts. Given an area one can determine the number of districts with 1, 2, ... neighbours. Given M marked districts one can count the number of their neighbours and from that the expected value of the number b of marked neighbours and the standard deviation [Cliff and Ord, 1981]. An area with M marked districts is considered clustered if the actual number of marked-neighbours pairs exceeds an appropriate significance boundary.

An above average number b of neighbourhoods between marked districts does not necessarily indicate a clustering; the reason may be that several marked districts have above average neighbours. Such a situation arrives easily in irregularly shaped areas such as counties in Germany. Therefore it makes sense to base a test conditionally on the number of neighbours of the marked districts. Making use of some slight approximations, one gets the following test.

Let $B := \sum B_j$ be the total number of neighbours of all districts, g_j the number of neighbours of the *marked* district j ($g_j = 0$ otherwise), b the total number of *marked* neighbours of *marked* districts, $G = \sum g_j$, $G_2 = \sum g_j^2$. b is the test statistic; the configuration of marked districts is considered not random if[1]

$$\left| b - \frac{G^2 - G_2}{B - \frac{G}{M}} \right| - \frac{1}{2} > ds \qquad \text{where}$$

$$\begin{aligned} s^2 &= 2B\left[\frac{\binom{M}{2}}{\binom{F}{2}} - 2\frac{\binom{M}{3}}{\binom{F}{3}} + \frac{\binom{M}{4}}{\binom{F}{4}}\right] \\ &\quad +4\sum B_j^2\left[\frac{\binom{M}{3}}{\binom{F}{3}} - \frac{\binom{M}{4}}{\binom{F}{4}}\right] \\ &\quad +B^2\left[\frac{\binom{M}{4}}{\binom{F}{4}} - \left(\frac{\binom{M}{2}}{\binom{F}{2}}\right)^2\right] \end{aligned}$$

Three isolated marked districts have no pairs of marked neighbours; a chain of three marked districts has two pairs; a triplet has three pairs. The more compact the marked districts are, the larger becomes our test statistic. The test needs no simulation experiments. It distinguishes less sharply than the first test between clusters (as defined here) and long chains of marked districts such as the circle around Paris in figure 1.

Comparing both tests of this section shows no uniform picture. Using real data, the neighbourhood count sometimes responds to concentrations in an area not really forming a cluster (resembling Swiss cheese). The cluster test is sometimes hampered by small-number effects (the distribution functions of both test statistics are step functions rather than continuous). The power of the tests depends markedly on the counter hypothesis, that is on the way to produce non-random configurations. Various simulations have been performed. Surprizingly, the cluster test tends to have a somewhat larger power for small portions of marked districts than the neighbourhood count test (around 10% larger), but often a considerably lower power for large portions. Incidentally, the conventional black-black-count test has sometimes a somewhat larger power than our neighbourhood count test, sometimes a markedly lower one.

The distribution of mumps départements in figure 1 is non significant with both tests, neither is the random assignment in figure 2.

---

[1] The formula given here is a recent improvement over that published in [Gebhardt, 1997b]. A new publication is planned [Gebhardt, 1998].

3 Test for scalar data

Finally we consider the case of scalar data, i. e. a real-valued variable. There exist already some procedures for searching clusters. Moran's I (see e. g. [Anselin, 1988]) tests on the existence of spatial non-randomness; essentially it computes the spatial autocorrelation, i. e. a global value for the correlation of a district with the districts in a vicinity (perhaps neighbouring districts, perhaps districts within a certain distance). A significant value of I does not tell where any clusters are and small clusters of extreme values in an otherwise rather random distribution may not show up. In addition, it does not distinguish between a clustering of high or of low values.

The statistics G_i and G_i^* by Getis and Ord [Getis and Ord, 1992] are designed to find districts whose neighbours are significantly above (or below) average. They are restricted to inherently positive variables, excluding therefore residuals from regressions, among others. One has to perform one test per district (with obvious consequences on the error probabilities) and finds predefined clusters, implicitly given by the weight matrix.

The basis of our test is a comparison between the mean value for a potential cluster and that for the remaining districts. However, there are many potential clusters of a given size in an area. For instance, the 94 French dépertements in figure 1 have 554 different clusters of size $k = 5$ and 7081 of size 8. So the expected value for the largest mean of a cluster is rather high.

The mathematical treatment of this situation is quite involved and requires a number of approximations, some rather crude, for the number of clusters of a given size in an area of F districts, for the tail of the normal distribution and others. For instance, an empirical approximation for the number of clusters of size k is $0.7 \cdot 2.5^k F$, which mostly holds in the areas studied with an error of 30% for $4 \leq k \leq 9$ except for small areas (less than about 50 districts) and for chess board arrangements (four districts at each corner).

Finally one arrives at a linear function of k, whose coefficients can be computed from F and a nominal significance level α. Let y_i be the values for the districts in the potential cluster, x_i all values (including those in the clusters). Then the distribution is considered not random if

$$k\frac{(\bar{y}-\bar{x})^2}{\sum(x_i-\bar{x})^2} - 1.79\frac{k}{F} > \frac{1}{F}\left(1.85\log\frac{F}{\alpha} - 8.86\right)$$

The right-hand boundary increases by $(1.85\log 10)/F = 4.26/F$ if the nominal error probability is multiplied by 0.1. With this test, clusters of different sizes can be compared; the cluster yielding the largest value (if it exceeds the boundary) is the result wanted. However, clusters with moderately smaller test values should also be considered; a measure for "moderately smaller" is the dependency of the boundary on α just mentioned.

The test is quite crude mainly for two reasons. In its derivation, one assumes the independence of the mean values of all potential clusters, which is clearly not given. It should be possible to adjust the test for this effect at least to a large extent. The other reason is the strong dependence on the distribution of the district values. A normal distribution has been assumed; simulations with different distributions yield boundaries that are sometimes off by more than the effect of a factor 10 in the nominal error probability. On the other hand, the deviation from the normal distribution may just be the result of unusually high values in a region, so it helps in identifying the anomaly.

The test has been used in simulations for small values of k, i. e. $k \leq 8$ and areas up to 171 districts. It seems that larger areas will behave in the same way, but for increasing k the test becomes increasingly stringent, an effect of the increasing dependence of the means for different clusters. Despite the crude approximations, clusters in the data show up quite clearly in a number of experiments with different areas.

Of course one cannot test all existing clusters even of moderate size; therefore one will apply a variant of beam search: one checks all triplets and keeps those with the largest values for the test statistic; then one tests successively only those clusters of size k that can be found by adding an overlapping triplet to the clusters of size $k-1$ and $k-2$, keeping always the best ones only.

The use of clusters formed by triplets rather than any connectivity regions serves to reduce the the number of potentially significant regions by one order of magnitude already for $k = 4$ and by a factor 50 to 100 (in the areas used in experiments) for $k = 10$; this permits to use smaller significance limits and speeds up the search algorithm.

An alternative test and competitor is Moran's I despite the deficiencies mentiond in the beginning; according to ous simulations, it is surprisingly robust against the distribution of the district values. So a possible, though costly, approach could be to use Moran's I for testing for non-randomness and then to use our procedure for finding the cluster.

Applied to the mumps data, the test statistic is highly significant due to the fact that the values are non-normal and the two highest values fall into the cluster. Using the logarithms of the data instead, the cluster is barely significant.

4 Example

As a second example let us look at figure 3. It shows the incidences of measles (cases per 100 000 inhabitants) in January 1995. The data are taken again from Le Réseau Sentinelles, based on the data reported regularly by 1% of the general practitioners.

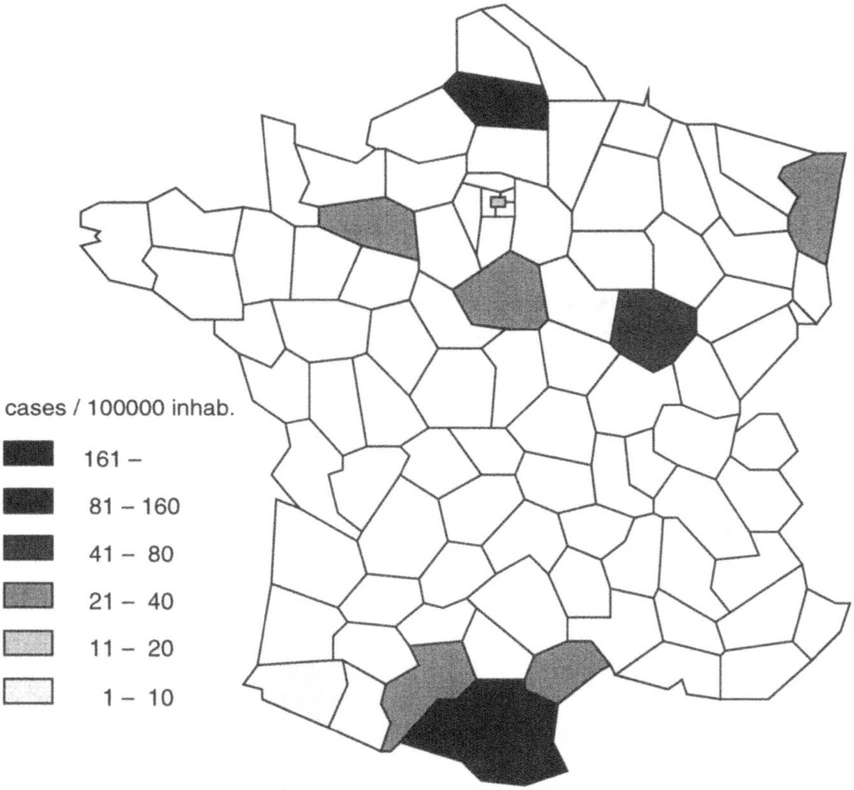

Figure 3: Measles, January 1995. Data taken from Le Réseau Sentinelles.

12 départements are marked, four of them form a cluster. This is just significant (about 2.2s). The neighbourhood test is not significant (1.5s); the marked districts have 55 neighbours, 12 of them are marked.

Using the numerical data, the situation is highly significant corresponding to a nominal error probability far below 10^{-10}%. The most outstanding cluster consists of three départements only; adding the fourth one, the nominal error probability increases by several powers of ten but still stays around 10^{-10}%.

The data are highly non-normal; so one can try again to use the logarithms. In this case, both clusters (three or four départements) yield approximately the same value corresponding to an error probability near $10^{-6}\%$, which still is highly significant.

Moran's I is approximately normally distributed; with the original data it yields 5.35, with the logarithms 3.43. Using binary data (marked vs. unmarked districts), Moran's I is insignificant (result: 1.85). So with these particular data, the tests show comparable results.

It should be mentioned that a significant clustering of the départements with measles cases is not the usual situation. In the other eleven months of 1995, there are between 2 and 20 marked districts; in two months, there is a (non-significant) cluster of three départements. The cluster test and the neighbourhood test are always non-significant; the same holds for Moran's I with the numerical data. Our test for numerical data is sometimes significant, mostly because of the extreme distribution (if almost all values are equal, in our case zero, then all triplets including the highest value are significant).

Several other experiments have been performed with other medical variables for the 94 départements and also with demographic variables for 171 German counties, with demographic data for 62 statistical districts of Bonn, with election data for 80 election districts in eastern Germany and with random data for these four areas as well as regular honeycomb arrangements of 37 to 169 hexagons.

References

[Anselin, 1988] Luc Anselin. *Spatial Econometrics: Methods and Models.* Kluwer, Dordrecht, 1988, 284 pages, ISBN 90-247-3735-4.

[Cliff and Ord, 1981] A. D. Cliff and J. K. Ord. *Spatial Processes: Models and Applications.* Pion, London, 1981, 266 pages, ISBN 0-85086-081-4.

[Gebhardt, 1996] Friedrich Gebhardt. Clusters in geographical distributions. Arbeitspapiere der GMD 1035, GMD, Sankt Augustin, 1996, 33 pages.

[Gebhardt, 1997a] Friedrich Gebhardt. Clusters in spatial area data. Arbeitspapiere der GMD 1068, GMD, Sankt Augustin, 1997, 21 pages.

[Gebhardt, 1997b] Friedrich Gebhardt. Finding spatial clusters. In Jan Komorowski and Jan Zytkow, editors, *Principles of Data Mining and Knowledge Discovery: First European Symposiun, PKDD '97, Trondheim, June 1997*, pages 277–287. Springer, Berlin, 1997, ISBN 3-540-63223-9.

[Gebhardt, 1998] Friedrich Gebhardt. Cluster test for geographical areas with binary data. *International Journal of Geographical Information Science*, 12:submitted, 1998.

[Getis and Ord, 1992] Arthur Getis and J. K. Ord. The analysis of spatial association by use of distance statistics. *Geographical Analysis*, 24:189–206, 1992.

[Toubiana and Flahault, 1998] Laurent Toubiana and Antoine Flahault. Monitoring the participation of sentinel general practitioners with health care workstation SITIE. In L. Gierl and M. Bull, editors, *GEOMED '97: proceedings*, in this volume. Teubner, Stuttgart, 1998.

[Toubiana *et al.*, 1995] Laurent Toubiana, Jean-Franҫis Vibert, and P. Garnerin. SITIE: a health care workstation integration architecture for epidemiologists. *Computers and Biomedical Research*, 28:100–115, 1995.

Index

Subject Index

GPSR Compliance

The European Union's (EU) General Product Safety Regulation (GPSR) is a set of rules that requires consumer products to be safe and our obligations to ensure this.

If you have any concerns about our products, you can contact us on ProductSafety@springernature.com

In case Publisher is established outside the EU, the EU authorized representative is:

Springer Nature Customer Service Center GmbH
Europaplatz 3
69115 Heidelberg, Germany

Batch number: 10726298

Printed by Printforce, the Netherlands